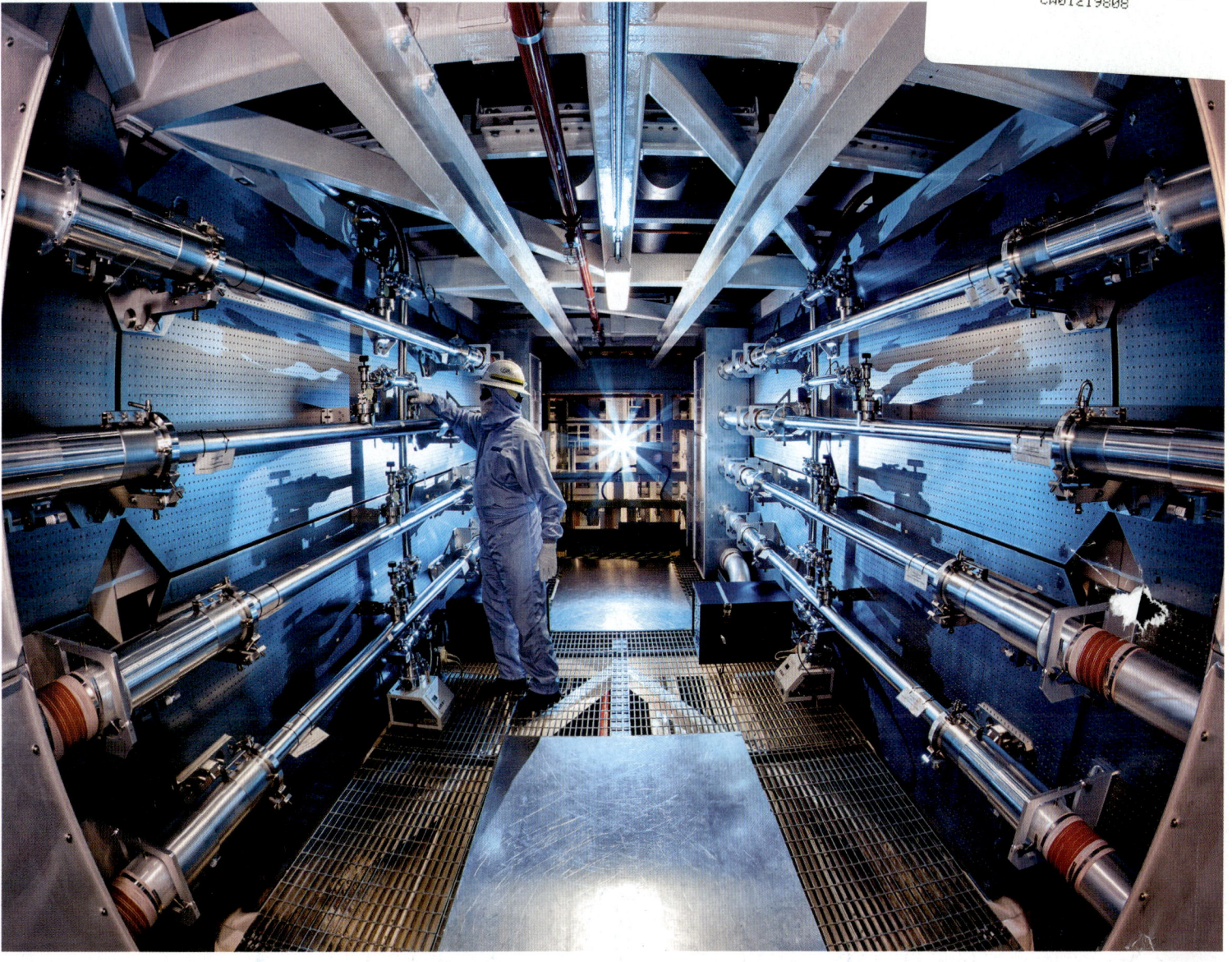

Welcome

Physics is a branch of science that wholly underpins the other natural sciences. It's a fundamental part of everything we do: from the simplest activity like hitting a ball with a bat, to slamming particles together at near the speed of light in multi-billion dollar accelerators, in order to help scientists answer some of the biggest questions about the universe. It's tough just to touch on the breadth of principles covered by this science in one book, but we've tried! Turn the page to discover how gravity keeps our feet on the ground, whether time travel is possible, and more...

FUTURE

UNDERSTANDING Physics

Future PLC Quay House, The Ambury, Bath, BA1 1UA

Understanding Physics Editorial
Editor **Ben Biggs**
Senior Art Editor **Duncan Crook**
Compiled by **Drew Sleep & Andy Downes**
Head of Art & Design **Greg Whitaker**
Editorial Director **Jon White**

Cover images
Future plc

Photography
All copyrights and trademarks are recognised and respected

Advertising
Media packs are available on request
Commercial Director **Clare Dove**

International
Head of Print Licensing **Rachel Shaw**
licensing@futurenet.com
www.futurecontenthub.com

Circulation
Head of Newstrade **Tim Mathers**

Production
Head of Production **Mark Constance**
Production Project Manager **Matthew Eglinton**
Advertising Production Manager **Joanne Crosby**
Digital Editions Controller **Jason Hudson**
Production Managers **Keely Miller, Nola Cokely, Vivienne Calvert, Fran Twentyman**

Printed in the UK

Distributed by Marketforce, 5 Churchill Place, Canary Wharf, London, E14 5HU
www.marketforce.co.uk – For enquiries, please email:
mfcommunications@futurenet.com

Understanding Physics Second Edition (HIB5956)
© 2024 Future Publishing Limited

We are committed to only using magazine paper which is derived from responsibly managed, certified forestry and chlorine-free manufacture. The paper in this bookazine was sourced and produced from sustainable managed forests, conforming to strict environmental and socioeconomic standards.

All contents © 2024 Future Publishing Limited or published under licence. All rights reserved. No part of this magazine may be used, stored, transmitted or reproduced in any way without the prior written permission of the publisher. Future Publishing Limited (company number 2008885) is registered in England and Wales. Registered office: Quay House, The Ambury, Bath BA1 1UA. All information contained in this publication is for information only and is, as far as we are aware, correct at the time of going to press. Future cannot accept any responsibility for errors or inaccuracies in such information. You are advised to contact manufacturers and retailers directly with regard to the price of products/services referred to in this publication. Apps and websites mentioned in this publication are not under our control. We are not responsible for their contents or any other changes or updates to them. This magazine is fully independent and not affiliated in any way with the companies mentioned herein.

FUTURE Connectors. Creators. Experience Makers.

Future plc is a public company quoted on the London Stock Exchange (symbol: FUTR)
www.futureplc.com

Chief Executive **Jon Steinberg**
Non-Executive Chairman **Richard Huntingford**
Chief Financial and Strategy Officer **Penny Ladkin-Brand**

Tel +44 (0)1225 442 244

Contents

EVERYDAY PHYSICS

10 How gravity works
The force that formed the stars and keeps our feet on the ground

14 Power of sound
Everything from thunder to a whisper is a series of sonic vibrations

18 The conservation of energy
Energy can't be created or destroyed, but what does that mean?

19 G-force explained
Intense acceleration in cars, fighter jets, rockets and roller coasters

20 The power of magnetism
What does it mean to be magnetised by this invisible force?

24 Centrifugal vs centripetal
There's a dizzying difference between these forces of circular motion

26 Refraction, rainbows and mirages
Amazing things happen when beams of light bend

28 Electricity basics
The science of resistance, inductance and capacitance

32 The science behind your swing
Find out why tennis is such a difficult sport to perfect

THE BIG THEORY

36 A beginner's guide to time travel
Why nothing in science says time travel is impossible

44 Newton's laws of motion
Three simple laws explain the effect of forces on the world

45 The general theory of relativity
Get to grips with Einstein's theory of the universe

46 What is string theory?
This strange idea could explain how the entire universe works

48 Matter vs antimatter
From space drives to bananas: all you need to know about the mysteries of antimatter

54 Testing Hawking's theories
Which of Stephen Hawking's ideas turned out to be right?

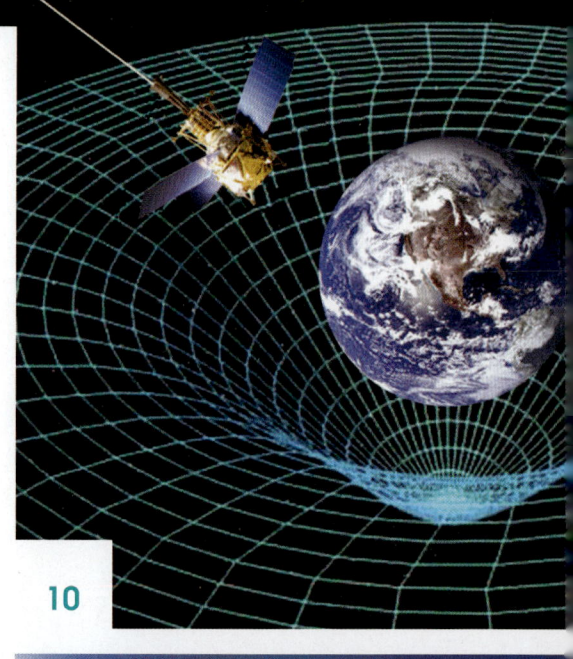

HIGH TECH

58 Nuclear Power
Today's nuclear power stations versus a nuclear fusion future

64 Small science
How have microscopes revealed the tiny world around us?

70 How superconductors work so efficiently
Turn down the temperature on these elements and materials to reveal their incredible superpowers

72 Inside an atom smasher
This particle accelerator is solving the mysteries of the universe

CONTENTS

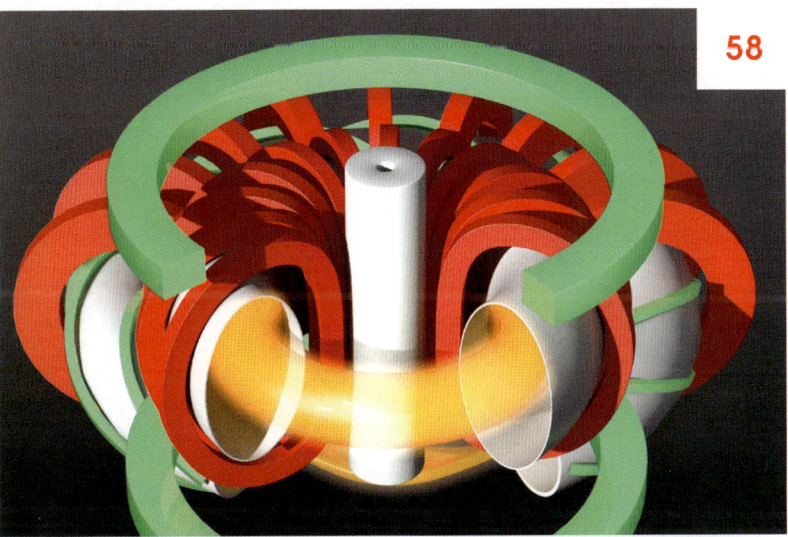

EXTREMES

80 Quantum power
The future of computing and how it will change your world

88 Extreme temperatures
Why leaving our usual range of temperatures makes materials do strange things

92 Deadly radiation
From dental X-rays to nuclear reactors: all you need to know about ionising radiation, its uses and hazards

100 The hidden universe
Dark matter and dark energy make up most of the universe, yet we can't see it. What is this strange stuff?

104 The power of atoms
Atoms are the ultimate construction kit, building everything from Venus de Milo to Venus the planet

THE UNIVERSE

112 Laws of the universe
Learn about the amazingly universal physics that governs the history, present and future of the cosmos

118 Why does Earth spin?
Find out why our planet and many others have rotational movement

120 The secrets of lightspeed
Can we ever overcome the universe's ultimate speed limit?

126 Measuring a galaxy's mass
How we work out the weight of galaxies in the universe

UNDERSTANDING PHYSICS

Everyday Physics

10 How gravity works
Unravel the force that formed the stars and keeps our feet on the ground

14 Power of sound
Everything from thunder to a whisper is a series of sonic vibrations

18 The conservation of energy
Energy can neither be created nor destroyed, but what does that mean?

19 Ge-force explained
Intense acceleration in cars, fighter jets, rockets and roller coasters

20 The power of magnetism
What does it mean to be magnetised by this invisible force?

24 Centrifugal vs centripetal
There's a dizzying difference between these forces of circular motion

26 Refraction, rainbows and mirages
Amazing things happen when beams of light bend

28 Electricity basics
The science of resistance, inductance and capacitance

32 The science behind your swing
Find out why tennis is such a difficult sport to perfect

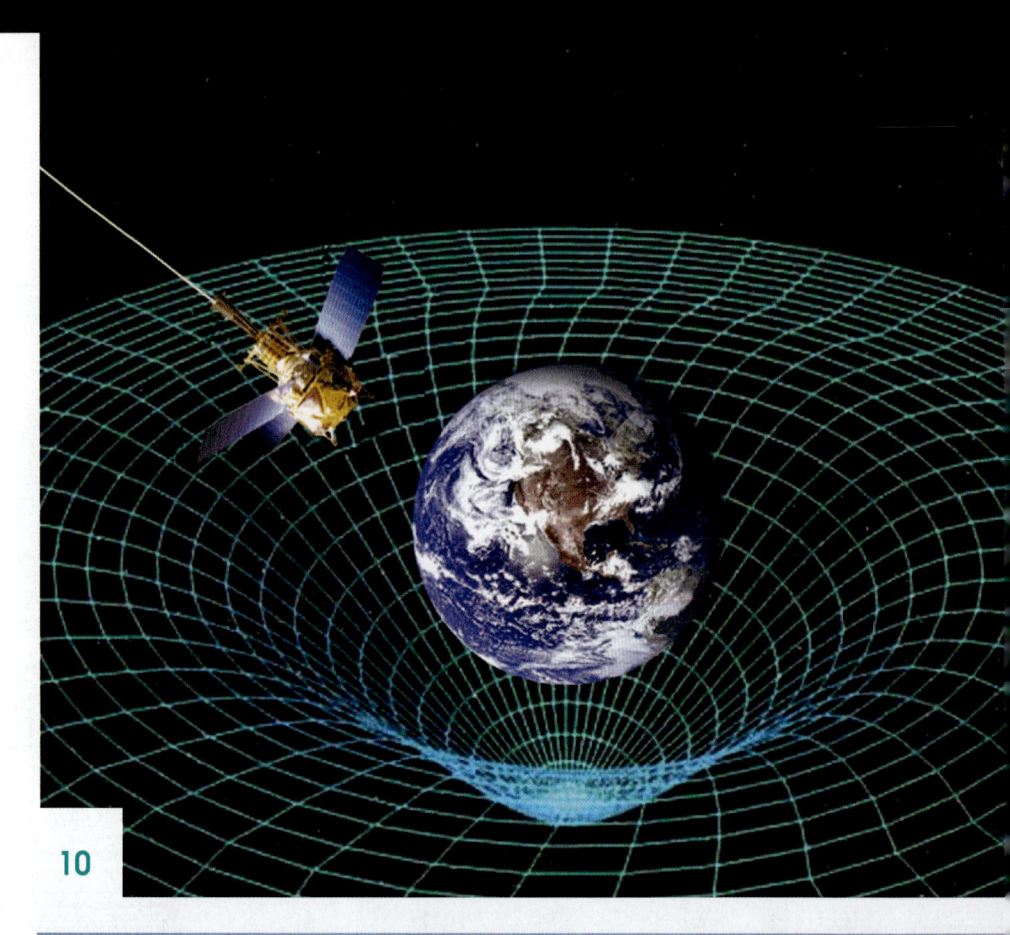

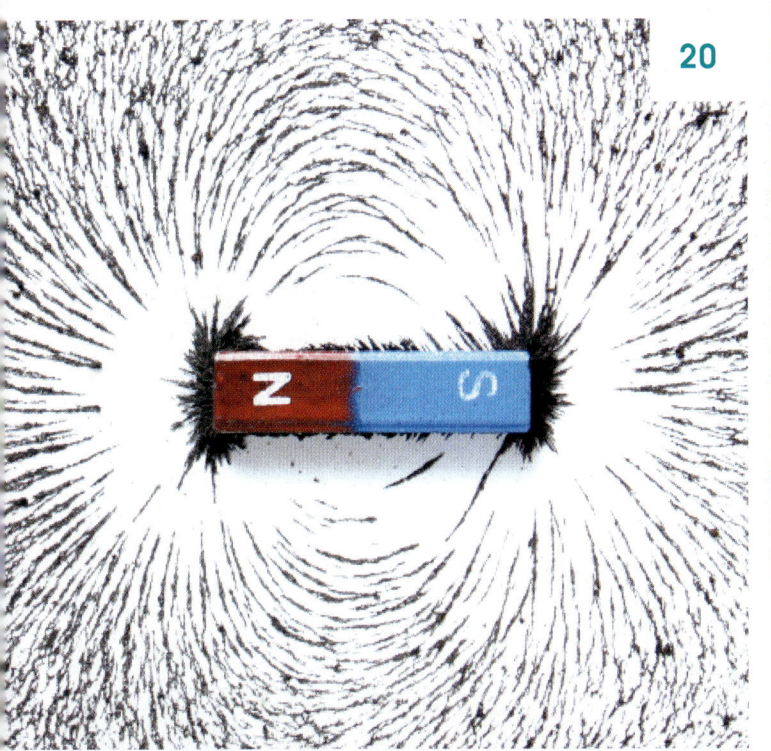

EVERYDAY PHYSICS

20

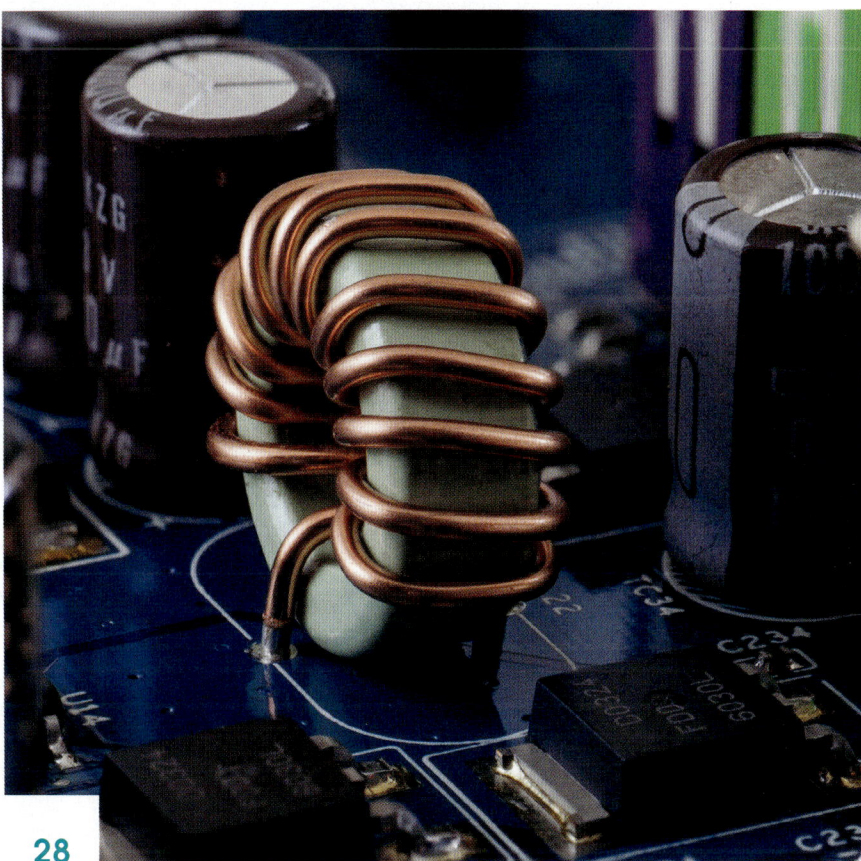

28

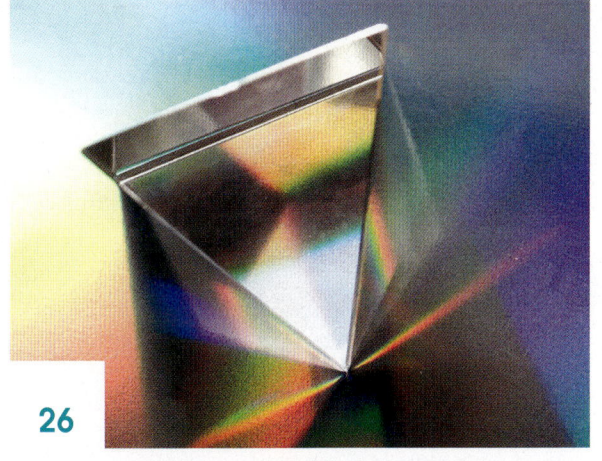

26

24  32

UNDERSTANDING PHYSICS

How gravity works

Unravel the mysterious force that formed the stars and keeps our feet on the ground

O f all Isaac Newton's revolutionary discoveries, perhaps none was more ambitious than unravelling the enigma of gravity. In the 1660s, Newton saw an apple fall to the ground and dared to ask, "Why?" Why doesn't the apple drift slowly upward? Why does water always seek the lowest place? Why does the Moon stay in orbit around Earth and not catapult into space? In his day, it was a question of near-religious significance.

Instead of meditating on divine mysteries, Newton drew up formulas. His law of universal gravitation, as presented in his 1687 treatise Principia, states that every particle of matter in the universe attracts every other particle of matter in the universe with a measurable force called gravity (named for the Latin 'gravitas', or weight). The strength of the gravitational force increases with mass and decreases with distance. In other words, the larger the object, the more gravity it exerts, and the closer you are to the object, the greater the pull.

Here is Newton's brilliantly simple formula for calculating the force of gravity between two objects, where m1 and m2 are the masses of the two objects, r is the distance between the two objects' centres of gravity, and G is the universal gravitational constant:

$$F = G \frac{m_1 m_2}{r^2}$$

Perhaps the most surprising thing about Newton's law is its universality. Though it can be difficult to conceive, not only is there a gravitational attraction between the apple and the Earth, but there's also a gravitational attraction between you and the apple. Essentially, any two objects that have mass – whether cosmically huge like a galaxy or infinitely small like an atom – exert a gravitational force on each other.

If that's true, though, why don't we swerve

EVERYDAY PHYSICS

HOW GRAVITY WORKS

Microgravity

Scientists use the orbiting ISS to conduct experiments in the weakened gravity 370 kilometres (230 miles) above the Earth's surface. In a microgravity environment, flames aren't drawn upward by convection currents. The steady, slow-burning flame of microgravity allows scientists to better understand the process of combustion both on our planet and beyond…

Gravity through the Solar System

As Newton theorised back in the 17th century, every particle of matter exerts a gravitational pull on every other particle of matter. If you concentrate a large amount of matter in one place, it will create a much greater gravitational pull than a loose smattering of particles. Mass is the measurement of how much matter is in a particular object. The greater the mass, then the more gravitational influence it will possess. Every planet, moon, star and galaxy in the universe has a different mass and therefore generates a unique gravitational pull. The mass of the Earth pulls a falling object toward the ground at a rate of 9.8m/s2 (32.2ft/s2). In contrast, the mass of the Sun is 333,000 times greater than the Earth. As a result, a falling object near the surface of the Sun would be pulled downward at a rate approaching 274m/s2 (899ft/s2), a staggering 28 times faster than on our planet.

Sun 28g
Mercury 0.378g
Venus 0.907g
Earth 1.00g
Mars 0.377g
Jupiter 2.36g
Saturn 0.916g
Uranus 0.889g
Neptune 1.12g
Pluto 0.059g

Gravity is the weakest force of nature, which is why a magnet can easily 'defy' it to pick up metal objects below

toward the street when a large truck passes, or get pinned to the base of a skyscraper? Because that 'big G' in Newton's equation is actually incredibly small – roughly 6.67 x 10-11 Newtons (square metres/kilograms); yes, the decimal point is 11 digits to the left. Unless the combined mass of the two objects is very, very large, the force of gravity between them is undetectable.

The Earth qualifies as a very, very large object with a mass of 5.97219 x 1024 kilograms (1.31664 x 1025 pounds). Not enormous on a cosmic scale but in comparison, your mass (not weight) is probably closer to 70 kilograms (154 pounds). If you plug the Earth's mass into Newton's equation as m1, your mass as m2, and then use the radius

How orbits work

There are over 900 satellites currently orbiting the Earth. But how do they stay in orbit without any engines? Satellites in orbit don't require power as they're really in controlled freefall. A satellite is launched into space in the nose of a rocket. That rocket must provide enough thrust to escape the surface gravity. Once in space, the satellite is released on a perpendicular trajectory. But instead of flying away from the planet, the satellite 'falls' into an elliptical orbit determined by the long-distance gravitational pull of the planet.

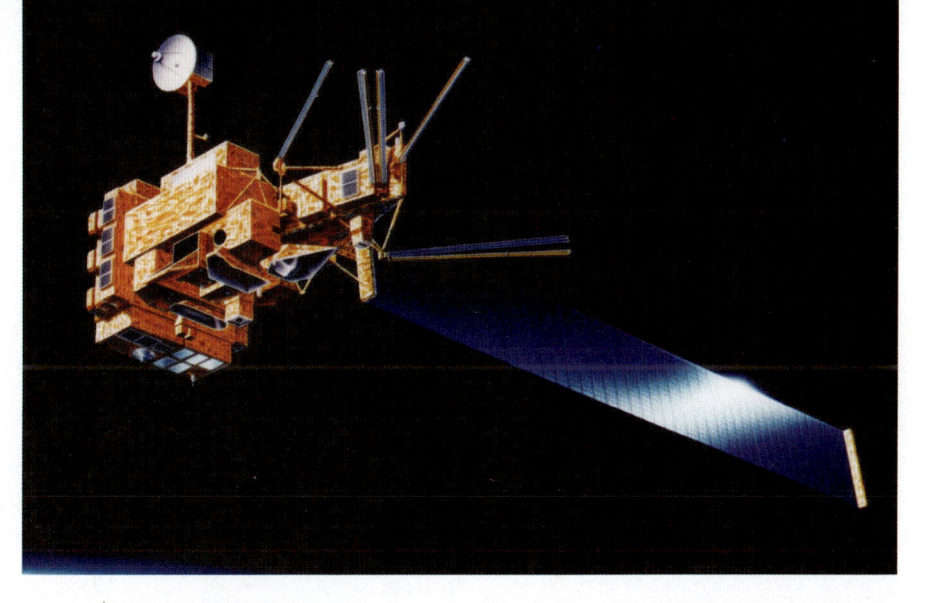

UNDERSTANDING PHYSICS

of Earth for r, you get an answer of 686 Newtons (154.2 pounds force).

That is the gravitational force between you and our planet – in other words, the force that your mass exerts through gravity, aka your weight on the surface of the Earth. If you were to run the same numbers at a jumbo jet's cruising altitude of around 12,200 metres (40,000 feet) above sea level, however, you would actually exert a whole two Newtons less, because there is a greater distance between your centre of gravity and the centre of the Earth.

Thanks to Newton's second law of motion, we know that force equals mass multiplied by acceleration (expressed as $f = ma$). Using Newton's gravity equation on page 40, we figured out the gravitational force between you and the Earth. Since we know the combined mass of you and the Earth, we can then solve the acceleration of gravity ($a = f/m$). The answer, $9.8 m/s^2$ ($32.2 ft/s^2$), is also known as 'little g'. Little g, like big G, is a constant, but it's only a constant for objects on or near the surface of the Earth. This means that little g on the Moon or near the Sun is a whole different story.

Little g is critical because it explains why objects fall to the Earth at a consistent rate, even when they are of wildly different masses. For instance, if you push a BMW Sedan and a bowling ball off the top of the Burj Khalifa hotel in Dubai –currently the tallest building in the world – they will both hit the ground at exactly the same time. The only exceptions are objects with low mass and a lot of surface area, like a feather or a parachute, which float down slowly as the result of upward drag. This wouldn't be the case, however, in an airless environment – for example, a laboratory vacuum or the surface of the Moon – where, believe it or not, the feather and the bowling ball would fall at precisely the same rate.

Notice that gravity is the force of attraction between two objects; that is, it's a two-way process. Not only are you attracted to Earth with a force of 686 Newtons (154.2 pounds force), but the Earth is attracted to you with an equal force. In fact, if you fall out of a tree and accelerate toward the Earth at $9.8 m/s^2$ ($32.2 ft/s^2$), the Earth also accelerates towards you. But that's impossible, right? The world doesn't jostle out of orbit every time some klutz tumbles out of a tree. The difference is in the rate of acceleration. If $a = f/m$ and f is 686 Newtons, then the rate of acceleration gets slower and slower as mass gets bigger and bigger. Yes, the Earth technically accelerates towards you and every other falling object, but that rate of acceleration is so tiny – and the Earth's inertia and momentum so great – that no wiggle is remotely detectable.

While Newton's universal law of gravitation gives us the physics to calculate the force and acceleration of gravity just about anywhere in the universe, it doesn't explain what gravity is and how it works at an atomic level.

Albert Einstein took up that challenge with his general theory of relativity, published in the early-20th century, which explained gravity as a curve in the space-time continuum. Beyond our three-dimensional universe, Einstein argues, is a fourth dimension of space and time. Objects with large masses, like planets, can warp the space-time dimension like a bowling ball on a trampoline. If you try to roll a marble across the trampoline, it will be drawn toward the bowling ball. The same is true for planets as they swirl in orbit around a huge celestial body like the Sun, or a cosmic beam of light that bends as it passes a black hole.

But even Einstein's revolutionary theory didn't explain the mechanism at work in gravity. What is it, exactly, that carries this force between two objects? Today, many physicists believe that the gravitational interaction is carried by undetectable, massless particles referred to as gravitons. Others talk of gravitational waves – the barely detectable shockwaves of gravitational force created by the collision of massive neutron stars, or the explosion of a supernova.

Despite the limits of our understanding, what began as an apple falling from a tree in the 17th century has led to remarkable insight into the mysterious forces that guide the universe. Gravity, the force that keeps our feet firmly on the ground and dictates global tides with the passing of the Moon, appears to be the same ancient force that bound together primordial cosmic elements to form the very first stars and galaxies, billions of years ago. If nothing else, it's something to mull over the next time you're falling out of a tall tree...

What goes up...

A fun way to experience weightlessness on Earth is to leave it momentarily. The flight of this motorbike follows a parabolic curve – the same path flown by NASA aircraft to ready astronauts for zero-gravity

Liftoff
The second the motorbike is airborne, the force of gravity drops to zero, giving the rider a weightless sensation.

Acceleration
Cruising across flat ground, the bike experiences normal gravity as it reaches a speed of around 104km/h (65mph).

Horizontal to vertical
When the bike hits the 45-degree ramp, it's forced upward against gravity, increasing the gravity force – G force – felt by the rider.

True weightlessness
At the top of the parabolic arc, the rider experiences the closest thing to true weightlessness on Earth, minus the drag of air resistance.

EVERYDAY PHYSICS

HOW GRAVITY WORKS

Gravity warps space and time
NASA's Gravity Probe B (pictured) is being used to test Einstein's general theory of relativity. He said that large masses, such as planets and other massive bodies, distort both space and time – as seen in the framework that represents space-time in this picture. More mass means more warping and greater gravity. In this artist's impression you can see how NASA's Gravity Probe B's super-sensitive gyroscope can detect the gravitational effect of Earth on both space and time, and the resulting distortion.

> "Little g is critical because it explains why objects fall to the Earth at a consistent rate"

Finding the centre of gravity
To calculate the acceleration of gravity, you need to know the distance between the centres of gravity of object one and object two. But how do you work out these centres of gravity? For a sphere like the Earth, it's easy. The centre of gravity is the exact centre of the sphere. The distance, then, between your centre of gravity and the Earth's centre of gravity is equal to the Earth's radius. For odder shapes like an apple or the human body, the centre of gravity is defined as the average centre of the object's mass. In practice, you can locate the centre of gravity of any object by finding its balancing point.

Landing
On the bike's landing, the rider experiences greater than normal gravity. An inclined landing ramp decreases the G force.

Coming down
With a take-off speed of 104km/h (65mph), a bike launched from a 45-degree ramp travels 22m (70ft) before gravity pulls it back to Earth.

A hammer vs a feather
Newton's law of universal gravitation states that an object with greater mass will exert a greater gravitational force. But force is not the same as acceleration. The question of which object lands first is a matter of acceleration. When you do the maths, you find that every object – regardless of its mass – has the same acceleration of gravity near the Earth's surface. Take a look:

$$a = f/m; \text{ or } a = (m \times 9.8m/s^2)/m; \text{ or } a = 9.8m/s^2$$

The only reason the feather falls slower on Earth is air resistance. In a perfect vacuum like space, in contrast, the feather and the hammer land at precisely the same time.

Measuring gravity

Thanks to Newton, gravity is a measurable force. Not coincidentally, the international standard unit of force is called a Newton (N). On Earth's surface, roughly 0.98N equals the downward force of gravity on 100 grams of mass. Likewise, one kilogram of mass exerts a downward force of 9.8N. To calculate the force of gravity, physicists use the formula f = ma (force = mass x acceleration). Since the acceleration of gravity is 9.8m/s2 on Earth – ie little g – we can easily calculate the Newton force of any mass. The average person's mass is 70 kilograms, which multiplied by 9.8 gives you 686N – the force by which gravity keeps us all securely grounded.

Newton was inspired by many scientists including Robert Boyle, Simon Stevin and René Descartes

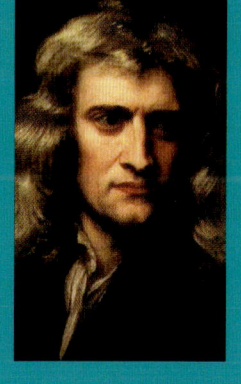

UNDERSTANDING PHYSICS

Power of sound

From a thunderclap to an infant's coo, the sounds of life are a symphony of sonic vibrations

If a tree falls in the wood and nobody is there to hear it, is it still a cliché? Yes, but it also makes a sound. When the tree strikes the ground, the force of the collision vibrates molecules in the trunk, branches and leaves of the tree, plus the dirt, rocks and plants on the ground. Those vibrations expand outwards in all directions at roughly 1,236kph (768 mph), moving as longitudinal waves of alternating high and low pressure. Whether or not an eardrum is in the vicinity to absorb those sound waves and send them to the brain for interpretation is beside the point. Sound is pure vibration.

Think about playing a drum in super-slow motion. Strike it with a stick and the drumhead momentarily compresses, then springs outwards. With each compression, the air around the drumhead drops in pressure, and with each expansion of the drumhead, there is an increase in air pressure. As the drumhead continues to vibrate, it sends out oscillating pulses of pressure that disturb the air molecules around the drum. High pressure makes air molecules momentarily compress together and low pressure pulls them back apart. That's why sound waves are called compression waves. They move through matter like the springs of a Slinky down the stairs.

Of course, when you hit a drum, the drumhead is not the only thing that vibrates. The wood and metal of the drum vibrate as well, all at slightly different frequencies. The emanating sound waves not only travel through air molecules, but through all molecules: solids, liquids and gasses. In fact, sound travels faster through solids than all other materials (15 times faster than in air), because the molecules are arranged in a tighter formation. The vibrations from the drum also bounce off the walls of the room, returning to the listener milliseconds later as reverberations. The result is a complex interaction of primary and secondary sound waves, harmonics and acoustics that we call music (or noise, depending on your age).

Picture sound waves like a series of classic sine waves. When graphing sound waves, the X-axis represents time and the Y-axis

EVERYDAY PHYSICS

POWER OF SOUND

The world's loudest noises
Decibel levels – up to 11

1 Rock concert
100 – 130dB
High-pressure sound waves (cymbals) can damage the eardrum at 120dB and the threshold for pain is listed at 130dB. Whether a rock concert is considered sound or 'noise' depends less on decibel levels than age.

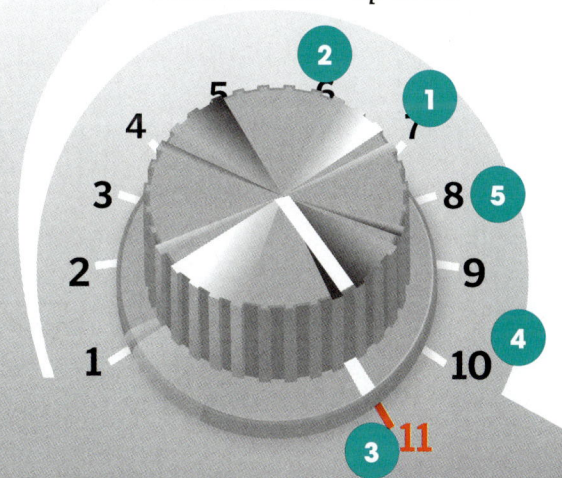

2 Lion
80 – 115dB
At a distance of one metre a fully grown male lion's roar can reach up to a rather noisy 115dB. In addition, at this intensity the roar can be heard at distance of some five miles (eight kilometres) away.

3 Krakatoa
180 – 200dB
Considered the loudest natural sound ever to occur in modern history, the 1883 explosion of the Krakatoa volcano was heard over 3,000 miles away. The explosion killed 40,000 people.

5 Sonic Boom
130 – 160dB
As an object approaches the speed of sound (jet, bullet, bull-whip), the sound waves in front of the object don't have time to get out of the way. Instead, they are compressed into a single, high-pressure wake or shock wave.

4 Space shuttle launch
150 – 190dB
The sound produced when NASA's Space Shuttle takes off is somewhere between 150 and 190 on the decibel scale. This level of loudness is enough to permanently damage a person's hearing.

---

measures air pressure. Above the centre line is positive air pressure and below is negative. Every characteristic of sound – volume, pitch, resonance – is determined by the shape of its wave. The taller the crest of the wave, the greater the amplitude. Since amplitude is a measurement of how much air pressure (or sound pressure) is created by the sound, we hear it as volume. Amplitude is expressed in decibels (dB). Zero decibels represents the lowest perceptible sound at around 20 micropascals of sound pressure. Normal conversation is around 60dB, and a pneumatic drill (jackhammer) is 100dB. New 'sound weapons' produce blasts of sound waves reaching 150dB, beyond the threshold of pain.

Frequency is a measurement of how many sound waves are created each second. Each cycle of the wave – crest to crest – per second equals one Hertz (Hz). We hear differences in frequency as pitch. Blue whales can produce extremely low-frequency tones – down to 10Hz – that can travel thousands of miles through the ocean. Low frequencies travel farther than high frequencies, which explains why you can hear the neighbour's kid's bass thumping when he's still three blocks away. A normal female voice can reach up to 1,100Hz, and anything above the threshold of human hearing (20,000Hz) is considered ultrasound. Bats and dolphins perform echolocation in the ultrasonic range and humans employ ultrasound equipment for medical and industrial imaging.

Interestingly, frequency can boost amplitude. Let's say a tuning fork vibrates at a natural frequency of 261.63Hz (middle C). Instead of striking it to make it vibrate, you can sing a middle C and it will start to vibrate. All materials have a resonant frequency that can be pushed into vibration by a matching tone. A tone sung at the perfect pitch can boost the amplitude of a crystal glass so far that it shatters.

The speed of sound

Sound waves travel at different speeds through different materials. In general, sound waves travel fastest through dense, stiff materials like iron and slowest through the air. That's because sound waves travel by temporarily disturbing molecules. One molecule bumps its neighbour, and so on until the strength of the sound wave dissipates. When molecules are tightly arranged in high-energy bonds, as in iron, they bump into each other with more frequency than in water or air, where molecules are spread further apart.

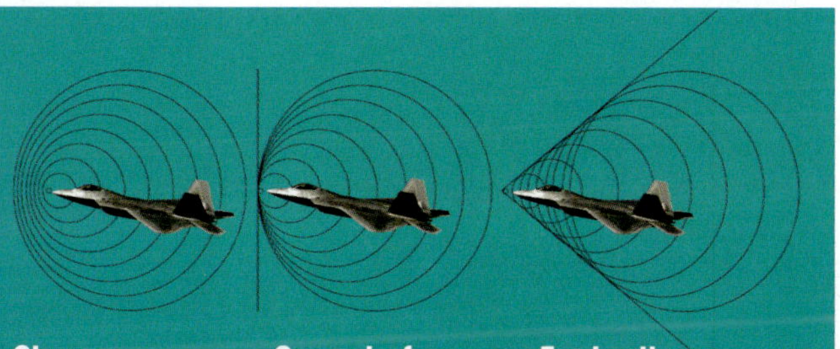

Slower than sound
As a jet flies through the air, its engine noise extends outwards at the speed of sound, like the ripples created when a stone is tossed into a pond.

Speed of sound
As the jet approaches 1200km/h (760mph), it overtakes the speed at which the engine noise extends outward from the plane, causing the sound waves to build up and compress.

Faster than sound
The jets breaks the 'sound barrier' when it's moving faster than the sound it is creating. The result is a wake of tightly compressed sound waves that are heard on the ground – not by the pilot – as a sonic boom.

UNDERSTANDING PHYSICS

Sound waves

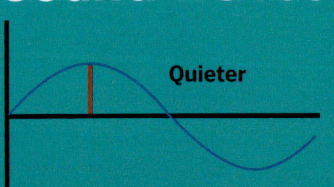

Quieter

Low amplitude
When graphing a sound wave, the X-axis represents time and the X-axis represents positive and negative air pressure. The positive height of a sound wave determines the amplitude or loudness of the sound. This wave, representing a quiet sound, has a low amplitude.

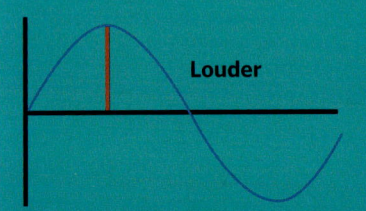

Louder

High amplitude
The taller the crest of the sound wave, the greater the positive air pressure, the greater the amplitude, and the louder the sound. Sounds are vibrations – oscillations of air pressure – so each positive pulse has a negative counterpart.

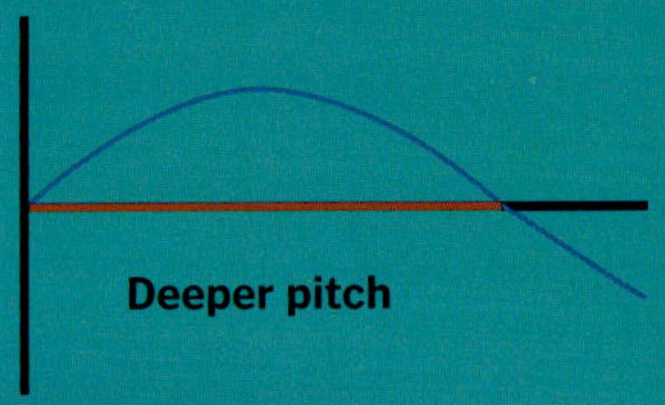

Deeper pitch

Low frequency
The pitch of a sound – what we perceive as high (treble) or low (bass) – is determined by the frequency of the sound wave. Frequency measures how many full wave cycles – crest to crest – occur every second. Each cycle equals one hertz (Hz). This long wave has a low frequency and produces a deep bass sound.

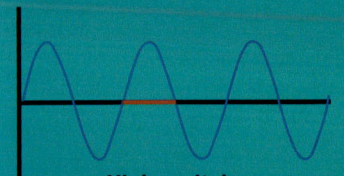

Higher pitch

High frequency
A female voice or a treble instrument like a flute produce sound waves of greater frequency. The faster the sound wave oscillates, the higher the pitch. The human ear can perceive sounds up to 20,000Hz, or 20,000 oscillations per second. Dogs can hear up to 60,000Hz.

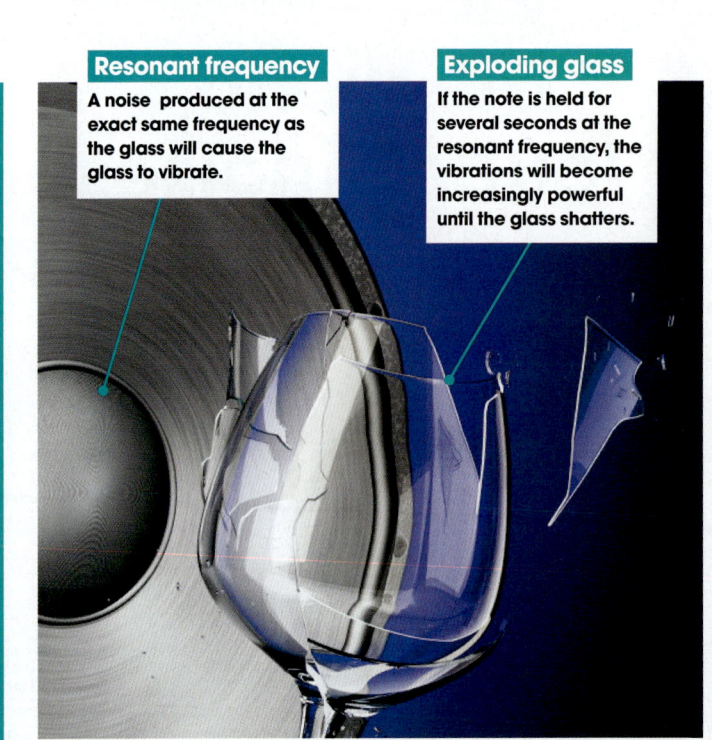

Resonant frequency
A noise produced at the exact same frequency as the glass will cause the glass to vibrate.

Exploding glass
If the note is held for several seconds at the resonant frequency, the vibrations will become increasingly powerful until the glass shatters.

Resonance explained

Every object vibrates at its own natural frequency. When you pluck the low E string on a guitar, it vibrates at an audible frequency that we perceive as music. When you start your car engine, the metal frame of the car vibrates, producing a distinctive sound. This natural frequency – dictated by the size, density, elasticity and material composition of the object – is also known as its resonant frequency. Interestingly, you can boost the amplitude, or intensity of a sound wave by matching its resonant frequency. Called 'sympathetic' resonance, it's how opera singers can shatter a crystal wine glass with their voice. By singing at the precise resonance frequency of the crystal, the glass begins to vibrate at larger and larger amplitudes until it shatters.

The Doppler effect

You're standing on the side of a road watching cars speed past. As a car approaches, the pitch of its engine noise appears to get higher and higher, but as it passes you, the pitch drops suddenly and continues to get lower. The same applies to the siren of a passing ambulance. To understand why, imagine sound waves emanating from the car like ripples on the surface of a pond. If the car is sitting still, the waves emanate at an even frequency in all directions. But as the car moves towards you, the waves arrive with increasing frequency, making it seem as if the pitch of the engine noise is rising. Conversely, as the car passes and drives away from you, it takes longer and longer for each sound wave to reach you, effectively lowering the frequency and therefore the pitch.

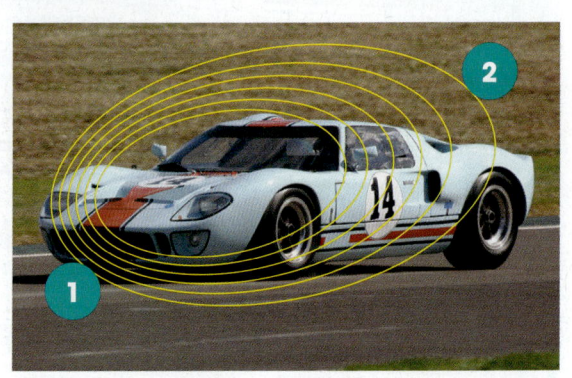

1 In front of the car
Sound waves ahead of the car bunch up and travel faster, making the waves shorter and the sound higher.

2 Behind the car
The sound waves spread out as the car passes by, causing a drop in pitch.

The lion's roar is the loudest of all the big cats. It can be heard for miles around

Doppler radar can detect wind shear, which poses a danger to aircraft during take off and landing

EVERYDAY PHYSICS

POWER OF SOUND

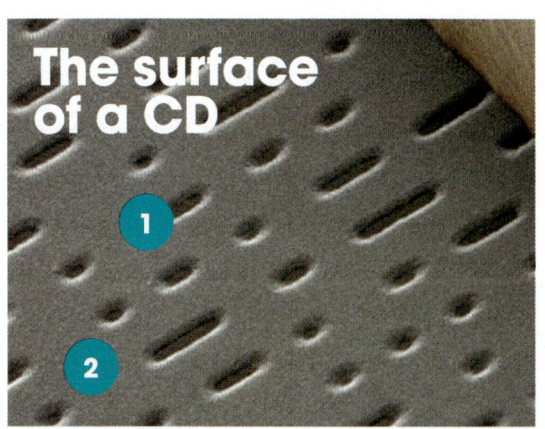

The surface of a CD

1 Pits
These microscopic pits create the digital code that represents the sound recording.

2 Lands
The spaces between the pits are known as the lands. Together these etchings are read by a laser reflected at the pitted surface.

Sound frequencies

How do these different noises measure up on the frequency scale?

Lesser horseshoe bat

Human vocal range

Blue whale song

93,000 – 111,000Hz
This tiny bat (200mm wingspan) employs one of the highest echolocation frequencies in nature, deep into the ultrasonic range. The highest audible frequency for humans is around 20,000Hz.

80 – 1,100Hz
These frequencies represent the normal combined male and female vocal range, but the world record for the highest note sung by a human is a high C on a piano, registering 4,186Hz. And the record holder is a man!

10 – 40Hz
Less a song than a moan, these super-low vocalisations are not only incredibly deep – the human ear can only perceive frequencies above 20Hz – but also incredibly loud, registering as high as 188dB at one metre.

KEY
Hz – Hertz: One wave oscillation per second
KHz – Kilohertz: 1,000 wave oscillations per second
MHz – Megahertz: 1,000,000 wave oscillations per second

Digital sound

If you record a melody played on a flute, the result is an analogue sound wave, a perfectly faithful graphical replication of the music. To get that recording on CD or MP3, you must use an analogue-to-digital converter, software that takes tens of thousands of samples per second of the recording (44,100/s for an audio CD, and up to 192,000/s for DVD audio) and assigns each one a value. When you graph these values, you end up with a nearly identical digital version of the analogue sound wave. Since digital information can be represented by binary code, it's easy to encode those ones and zeros on a CD or MP3 audio file. To play the music, a digital-to-analogue converter translates the code into electrical impulses that vibrate the diaphragm of the speakers or headphones, to faithfully re-create the original sound wave.

From analogue to CD

1 Analogue sound wave
Record yourself speaking and you can graph the continuously oscillating sound waves – the differences in frequency and amplitude that we hear as pitch and volume.

2 Digital sampling
Software can take tens of thousands of samples per second of that same recording and assign each one a value. Graph the values and you'll get an approximation of the original sound wave.

3 Digital encoding
Each of those values (from 0 to 65,536) is translated to binary code, clusters of ones and zeros that constitute the most basic language of all computer processors.

4 Laser burning
To imprint the binary code data onto a CD, a powerful laser etches 'pits' (125 nanometres deep) and leaves spaces between called 'lands'. These microscopic etchings are similar to the bumpy track of a needle on a record player.

5 Reading the CD
The pits become 'bumps' on the underside of the CD. A laser mounted in the CD player decodes the ones and zer0s into electrical impulses that, with help from an amplifier, vibrate speakers or headphones to produce the original sound waves.

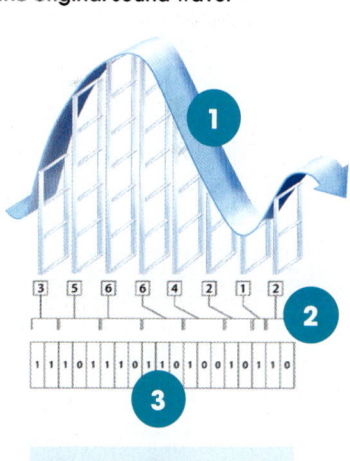

017

UNDERSTANDING PHYSICS

The conservation of energy

Energy can neither be created nor destroyed, but what does that mean?

Background

The conservation of energy is one of the most important concepts in physics. It states that in a system, the total amount of energy remains constant. Energy can be transformed from one form to another, from chemical to thermal for example, but it cannot be created or destroyed. This principle dictates much of our understanding of the world around us, and it forms one of the four laws of thermodynamics, the study of heat and energy.

In brief

In physics, energy describes the capacity for doing work. It comes in many different forms, which can be broadly divided into two groups: kinetic (movement) and potential (position). The example of a pendulum is often used to demonstrate conservation of energy in action. If you lift a ball on a string, it gains gravitational potential energy. When you let it go and it starts to swing down, its gravitational potential energy decreases, and its kinetic energy increases. As it passes the bottom of its arc and starts to swing upwards, it slows down; its kinetic energy decreases and its gravitational energy increases again. The energy isn't lost, it's just transferred from one type to another. With each swing a small amount of energy is also transferred as heat to the surrounding air, which is why the ball gradually slows down.

Summary

The total amount of energy in a system remains constant. Therefore energy cannot be created or destroyed, but it can be transferred from one form to another.

The law in action

Conservation of energy can be demonstrated by the swing of a pendulum

Minimum kinetic energy
At the top of the swing, the ball stops moving before it changes direction. It briefly has no kinetic energy.

Maximum potential energy
At the top of the swing, the ball is furthest from the ground, and has the most gravitational potential energy.

Maximum kinetic energy
At the bottom of the swing, the ball is moving at its fastest and has the most kinetic (motion) energy.

Friction
The ball slows down due to friction, but the energy isn't lost – it is transferred as heat energy to the particles in the air.

Minimum potential energy
When it reaches the bottom of the swing, the ball cannot get closer to the ground, so it has minimum gravitational potential energy.

Total energy
The total amount of energy in the pendulum system does not change.

Changing type
The type of energy that the pendulum has changes as it swings around.

Top of swing — Bottom of swing — Top of swing

KEY: Red = Gravitational potential energy   Blue = Kinetic energy

Julius Mayer

The law of conservation of energy was first described in the 19th century by Julius Mayer, a medical doctor from Germany. Mayer started his experiments as a child; he wanted to create a machine that could pump water around a water wheel using only the energy created by the wheel itself – essentially, a self-powering machine that generates energy from nothing. Try as he might, he could not find a solution. As an adult, he turned his attention to the energy produced by the human body. He used his observations to make the link between heat energy and mechanical energy, concluding that living things are just machines, and they too cannot create energy from nothing.

EVERYDAY PHYSICS

He's smiling now, but wait until he realises he left his parachute in the plane...

You spin me right round baby right round...

Acceleration can knock you out

If you are riding in a jet while it is making sharp turns, the blood in your head may rush out into your lower body. As the plane turns, all of the fluids in your body will act as if they were in a centrifuge, moving toward your feet, or whatever part of your body is on the outer edge of the turn. When that happens, your eyes will not get enough oxygen and you may experience a greyout, a sudden loss of colour vision, or a full blackout, temporary blindness. Accelerate harder and you will lose consciousness as blood retreats from your brain, depriving it of oxygen. Some people will experience these effects below 5g, but seasoned fighter pilots can take a bit more because they are very physically fit. They train themselves to resist the forces, and wear special suits that squeeze blood up into their heads.

Not a good time to have a blackout...

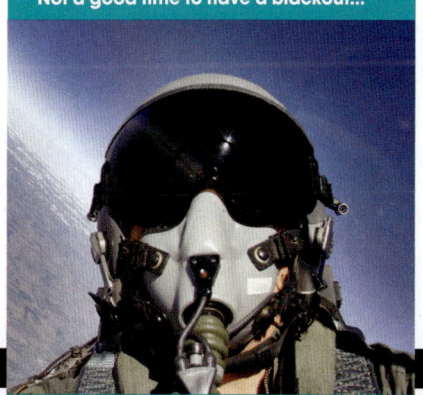

G-force explained

Intense acceleration in fast cars, fighter planes, rockets and roller coasters

When you're hurtling down the steel track of a roller coaster, it might seem that your stomach is climbing into your throat, and your eyes are squishing deep into your skull. Several forces are at play when you feel that way. Earth is constantly pulling down on every one of us. It has a great deal of mass, and that gives it a large gravitational field. And when we take a sharp turn on a fast ride, blast off in a rocket, or slam on the brakes, we're thrown around by forces far stronger than Earth's gravity. But why?

Engineers rate those experiences with numbers called g-forces, to explain how strong they are. One g is the amount of force that Earth's gravitational field exerts on your body when you are standing still on the ground. Every particle that makes up our planet is tugging on you simultaneously. Each one of those pulls is quite weak, but combined they are strong enough to keep your feet on the ground. Five g acceleration, something that race car drivers regularly experience, is five times as intense. Any time that an object changes its velocity faster than gravity can change it, the forces will be greater than one g. At zero g, you would feel weightless. And past 100g, you're almost certainly dead. Forces that intense can crush bones and squash organs.

Gravity is not the only source of g-forces. They take hold whenever a vehicle, like a car or a plane, suddenly changes its velocity. Speed up, slow down, or make a turn, and your velocity will change. The faster it happens, the more force you will experience.

Understanding g-forces

To find out how many gs you experienced during an intense acceleration, take your maximum speed, divide it by the time it took to hit that rate, and then divide by 9.81m/s2. The resulting number is how many gs you experienced.

Example: Put the pedal to the metal in a Bugatti Veyron and you will go from 0-100kph in 2.3 seconds.

100kph is 28m/s, 28 / 2.3 = 12m/s2, 12 / 9.8 = 1.2g

UNDERSTANDING PHYSICS

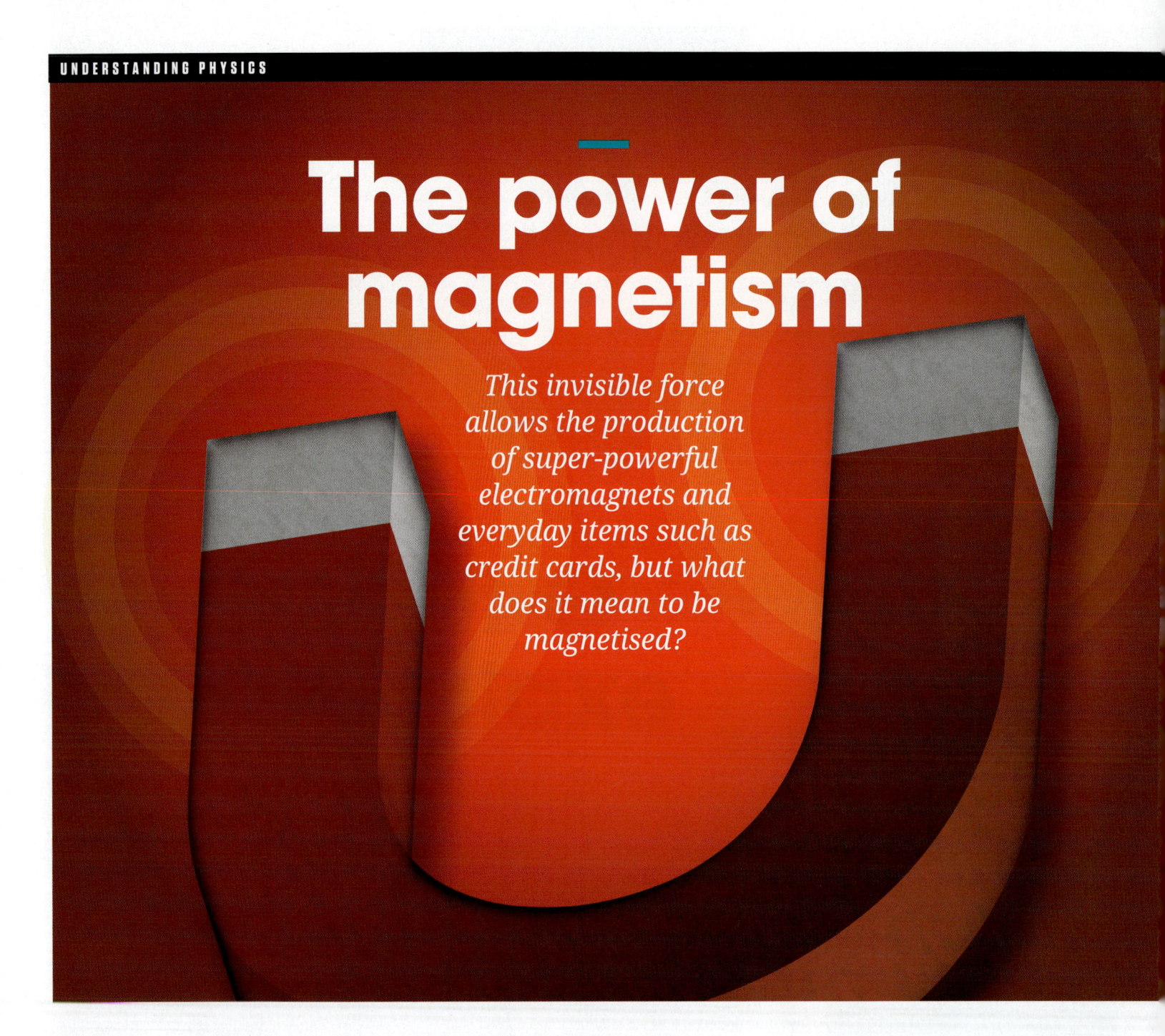

The power of magnetism

This invisible force allows the production of super-powerful electromagnets and everyday items such as credit cards, but what does it mean to be magnetised?

Magnetism is the force of nature responsible not only for our ability to live on a rock floating through space, but also for major technological achievements that have advanced the human race like never before. Our computers rely on them, our livelihood on Earth depends on their principles and our greatest science experiments use the most powerful magnets ever created by man. Were it not for magnetism we simply would not exist, and indeed without discovering the power of this fundamental force of nature, our life on Earth would bear no resemblance to what it is today.

Scientists over the years have employed magnetism in new and innovative ways, delving into realms of particle physics otherwise unexplored, but let's take a look at how basic magnets are made. It's fairly common knowledge that objects can be magnetised, making them stick to other magnetic objects, and we know that things such as a fridge or horseshoe magnet always have magnetism. To make permanent magnets like these, substances such as magnetite or neodymium are melted into an alloy and grounded into a powder. This powder can be moulded into any shape by compressing it with hundreds of pounds of pressure. A huge surge of electricity is then passed through it for a brief period of time to permanently magnetise it. Typically, a permanent magnet will lose about one per cent of its magnetism every ten years unless it is subjected to a strong magnetic or electric force, or kept at a low temperature.

Now let's take a look at the magnets themselves, and what's in and around them. Surrounding every magnet is a magnetised area known as a magnetic field that will exert a force, be it positive or negative, on an object placed within its influence. Every magnet also has two poles, a north and south. Two of the same poles will repel, while opposite poles attract. Inside and outside a magnet there are closed loops known as magnetic field lines, which travel through and around the magnet from the north to south pole. The closer together the field lines of this magnetic field are, the stronger it will be. This is why unlike poles attract – the magnetic forces are moving in the

EVERYDAY PHYSICS

THE POWER OF MAGNETISM

Magnetic atom

So what's the difference between the atoms of magnetic and non-magnetic elements? Well, the main difference is the appearance of unpaired electrons. Atoms that have all their electrons in pairs can't be magnetised, as the magnetic fields cancel each other out. However, atoms that can be magnetised have several unpaired electrons. All electrons are essentially tiny magnets, so when they are unpaired they can exert their own force – known as a magnetic moment – on the atom. When they combine with electrons in the other atoms, the element as a whole gains a north and south pole and becomes magnetised.

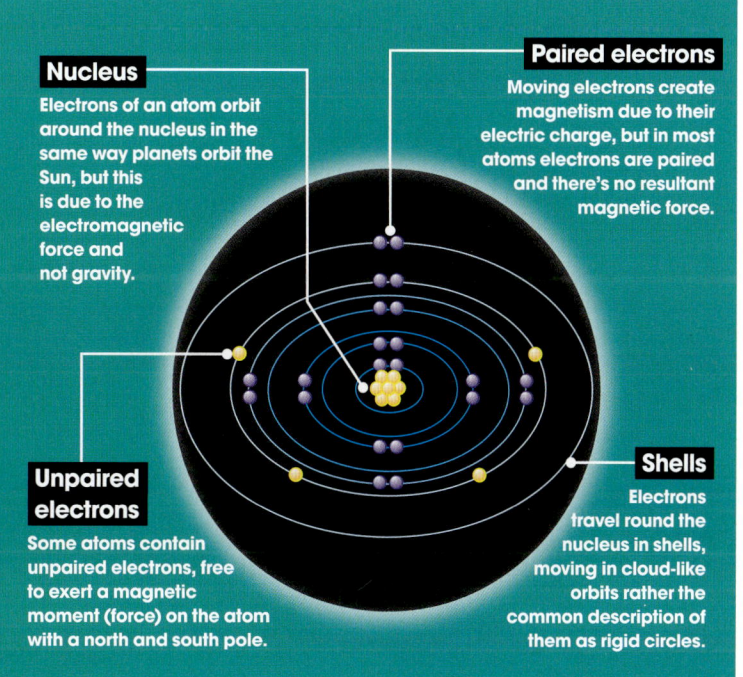

Nucleus
Electrons of an atom orbit around the nucleus in the same way planets orbit the Sun, but this is due to the electromagnetic force and not gravity.

Paired electrons
Moving electrons create magnetism due to their electric charge, but in most atoms electrons are paired and there's no resultant magnetic force.

Unpaired electrons
Some atoms contain unpaired electrons, free to exert a magnetic moment (force) on the atom with a north and south pole.

Shells
Electrons travel round the nucleus in shells, moving in cloud-like orbits rather the common description of them as rigid circles.

Inside a magnet

An object that can become magnetic is full of magnetic domains, chunks of about one quadrillion atoms. When the object is magnetised, the domains line up to and point in the direction of the magnetic field now present. This is why a magnetic object is sometimes stroked with a magnet to magnetise it. It aligns the domains in one direction, so that a magnetic field can flow around the material.

Unmagnetised
With no magnetism, the object does not have a north and south pole, so there is no magnetic field present to align the domains.

Scattered
When a substance that can be magnetic is unmagnetised its domains go in random directions, cancelling each other out.

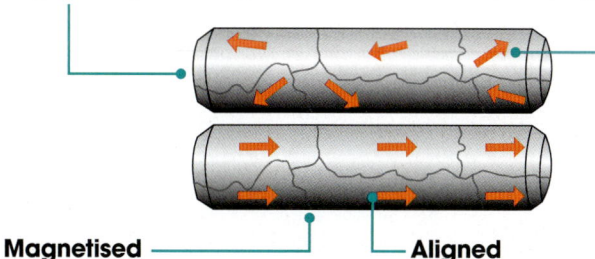

Magnetised
Introducing a magnet or electric current to the substance makes the domains all point in the same direction, with a magnetic field running from the north to south poles.

Aligned
When the domains are lined up, the substance as a whole becomes a magnet, with one end of it acting as a north pole and the other a south.

same direction, so the field lines leaving the south of one magnet have an easy route into the north of another, creating one larger magnet. Conversely, like poles repel as the forces are moving in opposite directions, hitting one another and pushing away. It's the same effect as other forces. If you push a revolving door while someone pushes from the other side, the door stays still and your forces repel. If you push in the same direction, however, the door swings round and you end up back at your starting point.

The defining feature of magnetic poles is that they always occur in pairs. Cut a bar magnet in half and a new north and south pole will instantly be created on each of the two new magnets. This is because each atom has its own north and south pole, which we will talk about later. However, the obvious question is why the poles are there in the first place.

Why do magnets have to have these field lines moving from north to south? The answer involves magnetic domains. It is best to picture a magnet as smaller magnet chunks put together. Each chunk (or domain) has its own north and south pole and again, as explained before, magnetic field lines travel from north to south. This means that all the domains stick together, with their forces concentrated in the same direction. They combine to make a larger magnet, exactly the same effect as when two magnets are stuck together. Each domain has about 1,000,000,000,000,000 (one quadrillion) atoms, while 6,000 domains are approximately equivalent to the size of a pinhead. Domains within a magnet are always aligned, but elements such as iron that can become magnetic initially have their domains pointing in random directions

Types of magnetism

Ferromagnetism
The strongest magnet in this list, a ferromagnet will retain its magnetism unless heated to a temperature known as the Curie point. Cooling it again will return its ferromagnetic properties. Every atom in a ferromagnetic material aligns when a magnetic field is applied. Horseshoe magnets are ferromagnets.

Ferrimagnetism
Ferrimagnets have a constant amount of magnetisation regardless of any applied magnetic field. Natural magnets like lodestones (magnetite) are ferrimagnets, containing iron and oxygen ions. Ferrimagnetism is caused by some of the atoms in a mineral aligning in parallel. It is different from ferromagnetism in that not every atom aligns.

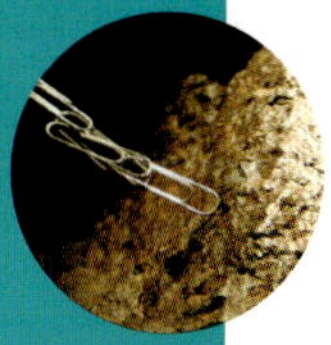

Antiferromagnetism
At low temps, the atoms in an antiferromagnet align in antiparallel. Applying a magnetic field to an antiferromagnet such as chromium will not magnetise it, as the atoms remain opposed. Heating to Néel temp (when paramagnetism can occur) will allow weak magnetism, but further heating will reverse this.

Paramagnetism
Paramagnets, such as magnesium and lithium, have a weak attraction to a magnetic field but don't retain any magnetism after. It's caused by at least one unpaired electron in the atoms of a material.

Diamagnetism
Gold, silver and many other elements in the periodic table are diamagnets. Their magnetic loops around the atoms oppose applied fields, so they repel magnets. All materials have some magnetism, but only those with a form of positive magnetism can cancel the negative effects caused by diamagnetism.

© Alchemist; Gregory F. Maxwell; Ryan Somma; Jeff Belmonte

021

UNDERSTANDING PHYSICS

when the iron is unmagnetised. They cancel each other out until a magnetic field or current is introduced, making them point in the same direction and magnetising the iron, which creates its own new magnetic field.

To really understand magnets, though, we need to get into exactly what is happening inside these domains. For that, we need to get right down into the atom. Let's take an iron atom, for example. Electrons circle the nucleus of an atom in cloud-like orbitals, commonly described as rigid shells (although in actuality, their motion is much more random). Each atom has a particular number of shells depending on how many protons and neutrons it has, while within each shell electrons orbit in pairs. Electrons are just like tiny magnets, each one having its own north and south pole. In their pairs, the electrons cancel out with one another so there is no overall magnetic force. In an atom such as that of iron, however, this is not the case. There are four electrons that are unpaired, which exert a magnetic force on the atom. When all the atoms are combined together and aligned, as we explained when talking about domains, the iron itself becomes magnetised and attracts other magnetic objects.

So we've snapped our magnet, broken it into chunks and subsequently examined the atoms of those tiny chunks. But can we go deeper? The answer to that is yes and no, as we delve into the unknown areas of quantum physics. The underlying principle of magnetism is that in the universe there are four fundamental forces of nature, being gravity, electromagnetism, the weak force and the strong force. Even smaller than atoms and electrons are fundamental particles known as quarks and leptons, which are responsible for these forces. Any force – such as gravity, magnetism, nuclear decay or friction – results from these fundamental forces. A force such as magnetism at this level is 'thrown' between particles on what are known as force carrier particles, pushing or pulling the other particles around accordingly.

Unfortunately, at this level magnetism goes into the realm of theoretical physics, entering areas of quantum physics that have not been explored in as much detail as particle physics. For now, this standard model of physics explains magnetism to a level that can only be furthered when we advance our understanding of quantum physics in the future.

Magnetic movement
The North Magnetic Pole moves up to 1mm a second because of changes in the core. In 2005 it was pinpointed just off Ellesmere Island in Canada, but is now moving towards Russia.

Earth's magnetic field
It's best to imagine the Earth as a bar magnet 12,400 miles (20,000km) long. The magnetic fields move around us like they would in a fridge magnet, but they also protect us from the universe. Compass needles always point to a magnet's south pole, so the Earth's geographical north pole is actually magnetically south.

Key: → wandering path of the magnetic north ⊕ rotational north pole

Effect
Charged particles from the Sun are deflected by the Earth's magnetic field, with some trapped in bands of radiation.

Tilt
The central 'bar magnet' of the Earth's magnetic field, the dipole, is tilted approximately 11° off the Earth's axis.

South is north
Magnetic fields always run from north to south, so when a compass points to the North Pole it is actually indicating southern magnetic polarity.

Cause
The magnetic field of any planet, including Earth, is the result of the circulation of electrically conducting material in the core, in our case molten iron.

Off-centred
The magnetic north and south poles do not draw a straight line through the centre of the Earth. In fact, they miss by several hundred miles.

Iron filings

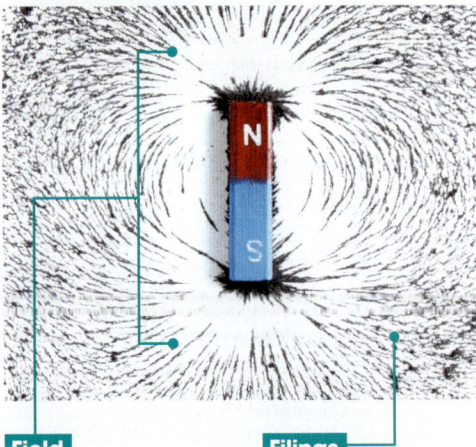

Field
By tapping the paper, the magnetic filings will align along the magnetic field lines of the magnet that run from its north to south pole.

Filings
By scattering tiny iron filings around a magnet on paper, it is possible to visibly see the magnetic field in action.

Electromagnets
One of the four fundamental forces in the universe, electromagnetism results from the interaction of electrically charged particles. Physicist Michael Faraday deduced that a changing magnetic field produces an electric field, while James Maxwell discovered that the reverse is also true: a changing electric field produces a magnetic field. This is the basis of how an electromagnet works.

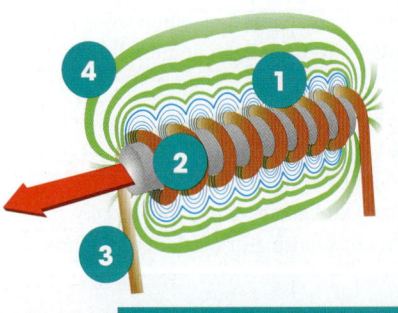

1 Electric fields
A wire wrapped around a magnetic core, such as iron, produces electric fields when a current runs through it, in turn creating a magnetic field.

2 Core
As discussed on the previous page, the domains within the core are unaligned until a magnetic field is introduced, created by a moving current in the coil.

3 Coil
The number of coils will increase the strength of the electromagnet because there is more current flowing in one direction, magnifying the force proportionally.

4 Magnetic field
The wire's magnetic field combines with the field of the core to produce a stronger field, with a larger current aligning more domains and increasing its strength.

EVERYDAY PHYSICS

THE POWER OF MAGNETISM

Magnets in your home
You'll be surprised at the number of magnets that are under your roof...

Doorbell
For a buzzer-style doorbell, pressing the button moves and releases a contact from an electromagnet to break and complete a circuit. A chiming doorbell, meanwhile, moves an iron core through an electromagnet coil and back when the button is pressed, hitting two chime bars in sequence.

Microwave
Inside a microwave oven is a magnetron, which contains magnets. Strong permanent magnets are mounted inside this tube. When electricity passes through the magnetron, the resultant electric and magnetic fields produce electromagnetic energy in the form of microwaves.

Vacuum cleaner
Electromagnetism is used here to produce the desired effect. A magnetically conducting material is inside the motor of the vacuum cleaner. When an electric current is introduced to a coil around the material, repulsive forces make the motor spin. The material loses its magnetism when the vacuum is turned off.

Computer
Like credit cards, the storage disks inside computers are coated with bits of iron. By changing the magnetic orientation of the iron, a pattern can be created to store a particular set of data. This pattern can be read by the computer and replicate the data on screen. The monitor itself uses magnets in the same way as an old cathode ray tube TV (see television).

Speakers
Using electromagnetism, most speakers contain a stationary magnet and a wire coil inside a semi-rigid membrane. When a current runs through the coil, the membrane rotates in and out because of the force between coil and magnet, creating vibrations that produce sound. Phone speakers use this same mechanism, only smaller.

Television
Most modern LCD or plasma TVs don't use magnets. However, older models use a cathode ray tube to fire electrons against the back of the screen. Coated in phosphor, parts of the screen glow when struck by the beam. Coils produce magnetic fields that move the beams horizontally and vertically to produce the desired picture.

Credit card
All credit cards have a black strip on them, known as a magnetic stripe. Inside, minuscule bits of iron are held in a plastic film. These can be magnetised in a north or south direction to store important data. When you swipe the card through a machine, the line of tiny magnets is read and information is obtained.

EMP
Buildings would survive; electronics wouldn't

An Electromagnetic Pulse (EMP) works by overwhelming electric circuits with an intense electromagnetic field. A non-nuclear EMP explodes a metal cylinder full of explosives inside a coil of wire, pushing out magnetic and electric fields that fry electric circuits. A nuclear EMP would explode a nuclear bomb in the atmosphere. The resultant gamma radiation would take in positive air molecules but push out negative electrons, sending a large electromagnetic field in all directions. A 10-megaton device detonated 200 miles (320km) above the centre of the United States would destroy every electronic device in the country but leave structures and life intact.

Solar blackout 2013?
Could the geomagnetic storm of 1859 be repeated?

In 1859, a great geomagnetic storm wiped out transmission cables and set fire to telegraph offices when the Sun went through a period of intense activity. Scientists at NASA have warned that a similar storm could occur in May 2013, when many more modern electrical components could be affected. The Sun's magnetic cycle peaks every 22 years, while every 11 years the number of solar flares hits a maximum. In May 2013 these events could combine and produce huge levels of radiation, potentially wiping out electric power on Earth for hours – or days.

UNDERSTANDING PHYSICS

Centrifugal vs centripetal

There's a dizzying difference between these forces of circular motion

Centripetal force keeps the carriage on a roller coaster moving in a circular motion around a loop

Centrifugal force is ubiquitous in our daily lives, but is it what we think it is? We experience it when we round a corner in a car or when an aeroplane banks into a turn. We see it in the spin cycle of a washing machine or when people ride on a carousel. One day it may even provide artificial gravity for spaceships and space stations. But centrifugal force is often confused with its counterpart, centripetal force, because they are so closely related – essentially two sides of the same coin.

Centripetal force is defined as the force necessary to keep an object moving in a curved path which is directed inward towards the centre of rotation, while centrifugal force is defined as the apparent force felt by an object moving in a curved path that acts outwardly away from the centre of rotation. While centripetal force is an actual force, centrifugal force is defined as an apparent force. In other words, when twirling a mass on a string, the string exerts an inward centripetal force on the mass, while mass appears to exert an outward centrifugal force on the string.

Both centrifugal and centripetal are the same force, but experienced in different directions.

If you are observing a rotating system from the outside, you see an inward centripetal force acting to constrain the rotating body to a circular path. However, if you are part of the rotating system, you experience an apparent centrifugal force pushing you away from the centre of the rotating circle, even though what you are actually feeling is the inward centripetal force, which is keeping you from flying off on a tangent.

This apparent outward force is described by Newton's laws of motion. Newton's first law states that a body at rest will remain at rest, and a body in motion will remain in motion unless it is acted upon by an external force. If a massive body is moving through space in a straight line, its inertia will cause it to continue in a straight line unless an outside force causes it to speed up, slow down or change direction. In order for it to follow a circular path without changing speed, a continuous centripetal force must be applied at a right angle to its path.

Newton's third law states that for every action, there is an equal and opposite reaction. Just as gravity causes you to exert a force on the ground, the ground appears to exert an equal and opposite force on your feet. When you are in an accelerating car, the seat exerts a forward force on you just as you appear to exert a backward force on the seat. In the case of a rotating system, the centripetal force pulls the mass inward to follow a curved path, while the mass appears to push outward due to its inertia. In each of these cases, there is only one real force being applied, while the other is only an apparent force.

Hammer throwers utilise centripetal acceleration to launch metal balls

EVERYDAY PHYSICS

CENTRIFUGAL VS CENTRIPETAL

Going in circles
Here's how these forces keep things rotating

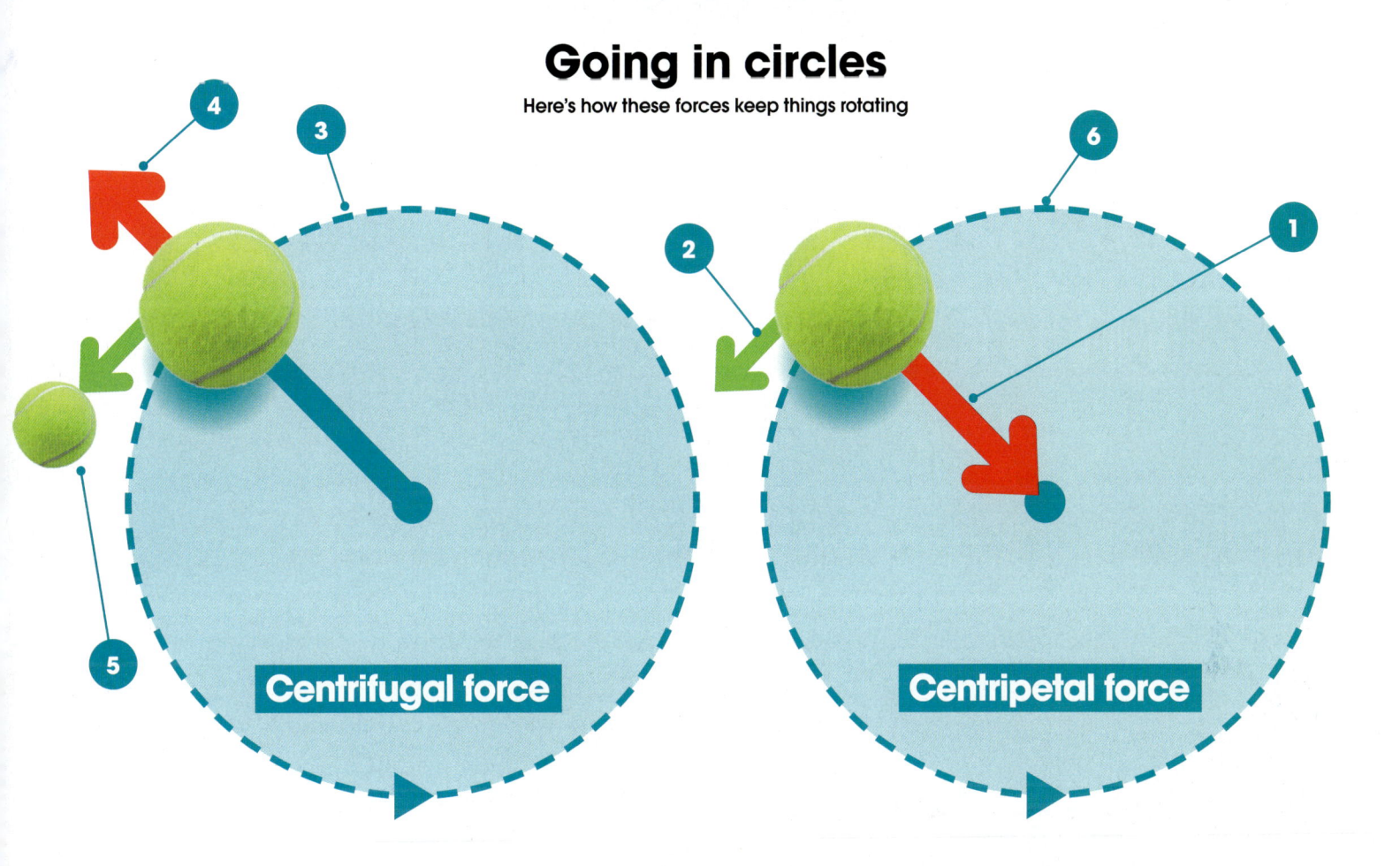

Centrifugal force

Centripetal force

1 Centripetal force
This is a force that pulls an object towards the centre of a circle.

2 Velocity
When an object's velocity is perpendicular to the direction a force is applied, such as tension on a rope, it will move in a circle.

3 Acceleration
The centripetal force increases if an object is moved closer to the centre of the circle.

4 Centrifugal force
This is the outward force generated when an object is being rotated.

5 Tangential direction
If the force keeping an object in rotation is broken, such as cutting the string that holds a spinning ball, the object will fly off in a straight line, following the tangential direction.

6 Inertia
According to Newton's first law, an object will constantly move in a straight line unless acted upon by another force, such as the tension of a spring, friction on a road or the pull of gravity.

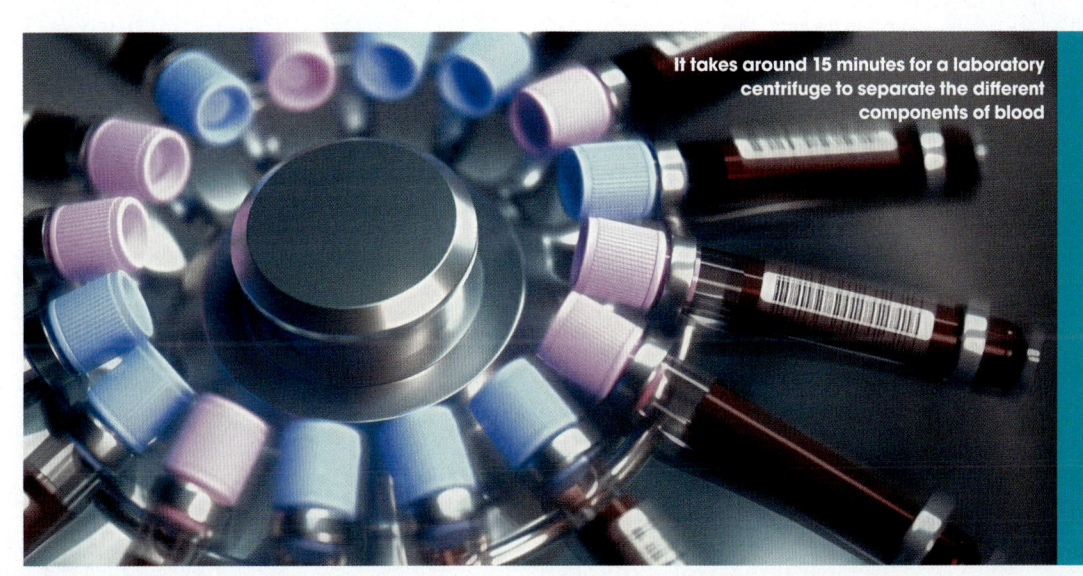

It takes around 15 minutes for a laboratory centrifuge to separate the different components of blood

Spun blood
Centripetal force is used in a laboratory centrifuge to accelerate the precipitation of particles suspended in liquid. One common use of this technology is for preparing blood samples for analysis. Under the normal force of gravity, thermal motion causes continuous mixing, which prevents blood cells from settling out of a whole blood sample. However, a typical centrifuge can achieve accelerations that are 600 to 2,000 times that of normal gravity. This forces the heavy red blood cells to settle at the bottom and stratifies the various components of the solution into layers according to their density.

UNDERSTANDING PHYSICS

Refraction, rainbows and mirages

Amazing things happen when beams of light bend

Glass prisms reveal the rainbow of colours hidden in white light

E veryone knows that nothing moves faster than the speed of light, but light doesn't always travel at its limit. It only reaches the dizzying speed of 299,792,458 metres per second in a vacuum. When light travels through any other material, be that a gaseous atmosphere or a glass of water, it interacts with atoms, and that slows it down.

If a beam of light hits a new material head-on, the wavelengths bunch up. They get closer together and the whole beam moves more slowly, but the light keeps moving in a straight line. When a beam hits a new material at an angle, something special happens. The part of the beam that hits first slows down first, and the light starts to bend. This plays tricks on our eyes, creating kinks in our drinking straws and puddles of water on dry desert floors.

Imagine a row of soldiers marching at an angle towards a line. Each soldier will slow down when they reach the line, but they don't all get there at the same time. When the first soldier arrives and adjusts their speed, the others carry on marching, causing the line to become staggered. The same thing happens when light hits a new material at an angle.

The amount the light bends depends on the 'refractive index' of the material it's moving through. This compares the speed of light in the material to the speed of light in a vacuum. For example, a refractive index of 1.5 means that light travels 1.5 times faster in a vacuum than in the material.

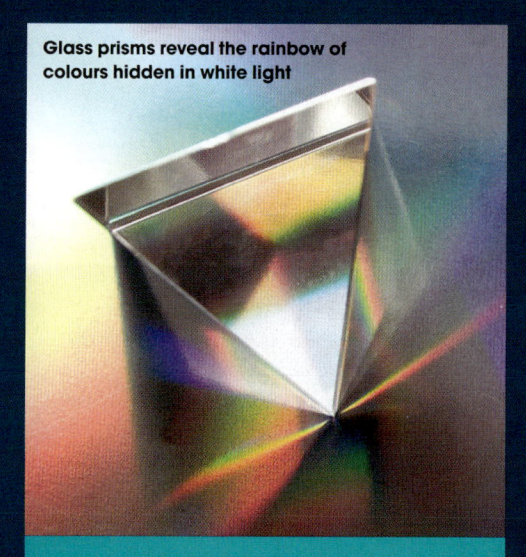

Newton's rainbows

It's been more than 300 years since Isaac Newton first tried his prism experiment, but the results are as dazzling as ever. At the time, people thought that colours were mixtures of light and darkness, and that white light was pure. Newton changed how we think about colour by placing a glass prism in a shaft of sunlight. When the light hit the prism at an angle, refraction separated the rays into a rainbow. The colours were always in the same order: red, orange, yellow, green, blue, indigo and violet. To prove that it wasn't the prism creating the colours, Newton put a second prism into the rainbow. Refraction bent the split beam back into a single stream of white light, demonstrating once and for all that white light contains all the colours.

Angle of incidence
When light hits the prism at an angle, parts of the beam slow down before others do.

Making rainbows
Recreate Newton's famous experiment with a beam of light and a set of prisms

Dispersion
Violet travels more slowly through the glass than red, spreading the light into a rainbow.

Bending light
The refractive index of each colour is slightly different in glass, ranging from 1.53 for violet to 1.51 for red.

White light
The experiment begins with a ray of white light. Newton created his by making a small hole in a window shutter.

"Refraction plays tricks on our eyes, creating kinks in our drinking straws"

EVERYDAY PHYSICS

REFRACTION, RAINBOWS AND MIRAGES

5 facts about mirages

1 Superior mirage
When warm air sits above cold air, the light bends downwards. This makes objects appear taller than they actually are and allows us to see things beyond the horizon.

2 Inferior mirage
When cold air sits above warm air, the light bends upwards, making the sky appear in puddles on the floor and creating classic desert mirages.

3 Late mirage
When a pocket of warm air blows over cold air, it's called a temperature inversion. When it happens above your eyeline, a strip of the Sun can seem to disappear.

4 Mock mirage
The effect of temperature inversions can change depending on their height. When they happen below your eyeline, they make wobbly horizontal slices through the sunset.

5 Fata Morgana
These complex mirages happen when there are alternating layers of hot and cold air. Also known as 'floating castle' mirages, they make objects look like they're levitating.

Trick of the light
Our brains expect light to move in straight lines. Here's what happens when it doesn't

Mirage
The brain thinks the light has travelled in a straight line, making the object appear much taller than it really is.

Warm air
Air expands as it heats up, making the gas less dense. This lowers its refractive index, letting light move faster.

Boundary
Light slows down as it moves from warm to cool air, and again from cool to cold.

Cold air
The gas molecules in cold air are closer together. This gives it a higher refractive index, making light move more slowly.

Reversal
Refraction also works in reverse. When rainbow light hits an opposite prism, it bends back to form white light.

Glass prism
Light slows down as it hits because of the change in refractive index: air is 1.0, glass around 1.5.

The rare 'green flash' mirage happens just before the Sun sets

The 'green flash' mirage

The tiny differences in the way the atmosphere refracts different colours of light aren't that noticeable during the day. But as the Sun sets, the effects can be dramatic. When the Sun drops below the horizon, a vibrant 'green flash' mirage can appear in its place. This rare afterimage happens because red light curves less than green light as it moves through the air. After the Sun sets, the red light rapidly disappears over the horizon. But if conditions are right, green light can continue to curve around the Earth for a few beautiful moments.

027

UNDERSTANDING PHYSICS

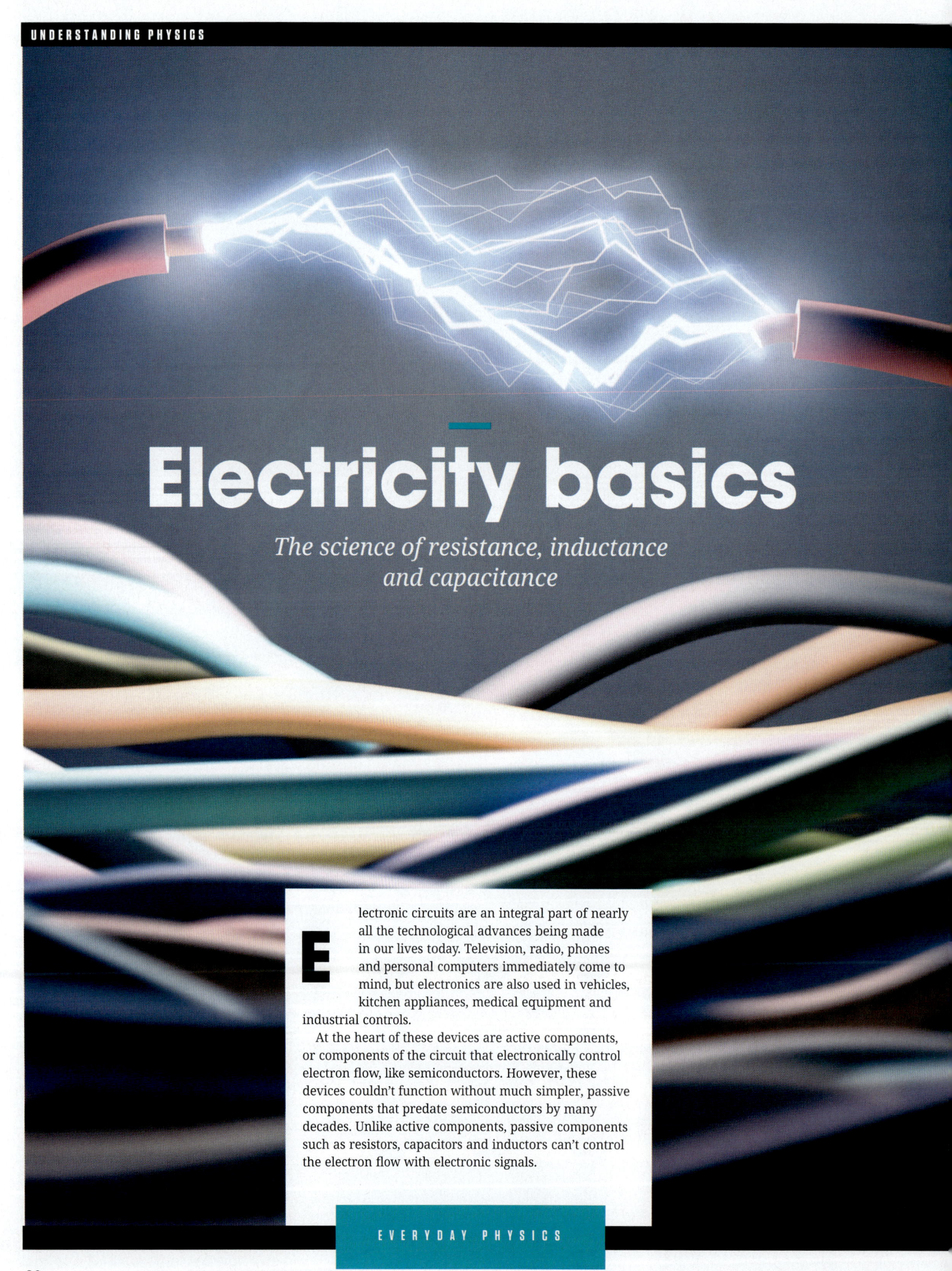

Electricity basics

The science of resistance, inductance and capacitance

Electronic circuits are an integral part of nearly all the technological advances being made in our lives today. Television, radio, phones and personal computers immediately come to mind, but electronics are also used in vehicles, kitchen appliances, medical equipment and industrial controls.

At the heart of these devices are active components, or components of the circuit that electronically control electron flow, like semiconductors. However, these devices couldn't function without much simpler, passive components that predate semiconductors by many decades. Unlike active components, passive components such as resistors, capacitors and inductors can't control the electron flow with electronic signals.

EVERYDAY PHYSICS

ELECTRICITY BASICS

Resistors are used in circuits to reduce current flow

German professor Georg Simon Ohm discovered the relationship between resistance and electric currents

Resistance

As the name implies, a resistor is an electronic component that resists the flow of electric current in a circuit. In metals such as silver or copper, which have high electrical conductivity and therefore low resistivity, electrons are able to skip freely from one atom to the next, with little resistance.

The electrical resistance of a circuit component is defined as the ratio of the applied voltage to the electric current that flows through it. The standard unit for resistance is the ohm, which is named after German physicist Georg Simon Ohm. This is defined as the resistance in a circuit with a current of one ampere at one volt. Resistance can be calculated using Ohm's law, which states that resistance equals voltage divided by current.

Resistors are generally classified as either fixed or variable. Fixed-value resistors are simple passive components that always have the same resistance within their prescribed current and voltage limits. They are available in a wide range of resistance values, from less than one ohm to several million ohms. Variable resistors are simple electromechanical devices, such as volume controls and dimmer switches, which change the effective length or effective temperature of a resistor when you turn a knob or move a slide control.

"Electrons are able to skip freely"

Electrical flow
The flow of electricity in a circuit is carried by passing charged electrons through a conductive material.

Resistance
The longer and thinner the copper, the slower the electrons pass through the resistor, lowering the voltage needed to power the circuit.

Wire
Inside some circuit resistors are many loops of wire-wound copper for the electrons to travel along.

Conversion
Resistors convert the energy of the electrical voltage into other forms of energy, such as thermal.

Creating resistance
How resistors slow down the flow of electrons in a circuit

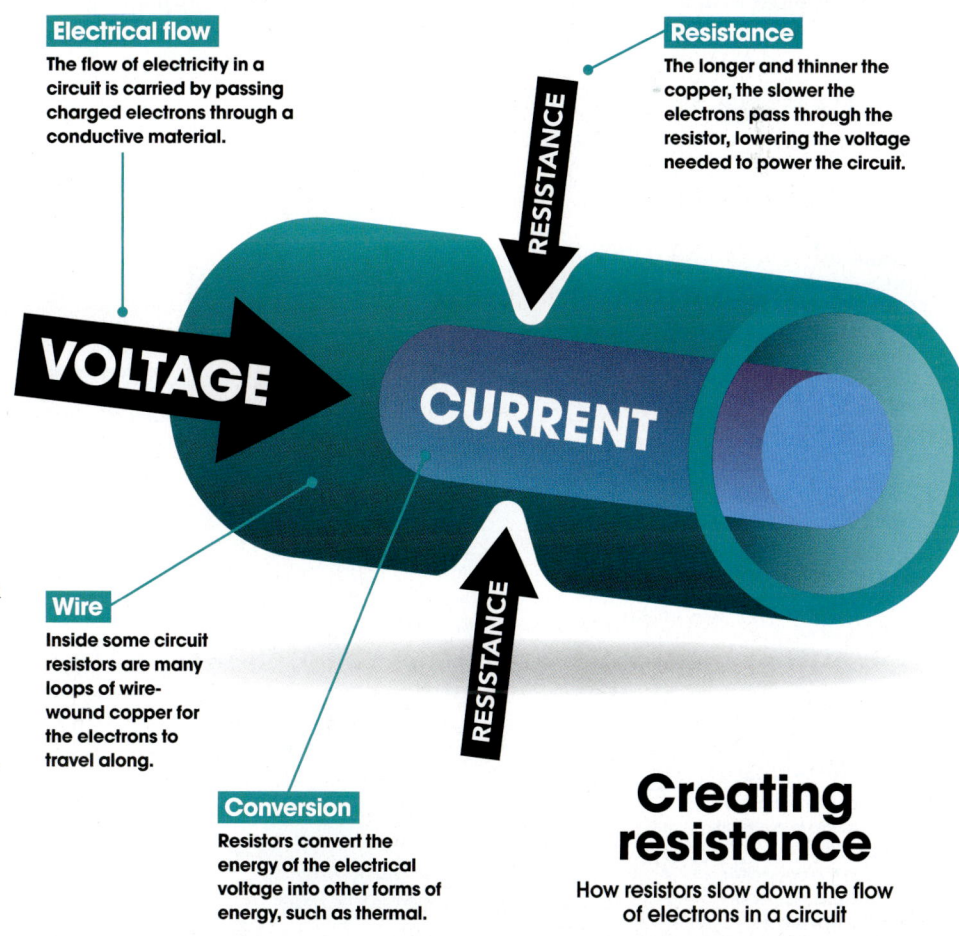

UNDERSTANDING PHYSICS

Capacitance

Capacitance is the ability of a device to store electric charge; the component that stores charge is called a capacitor. The simplest consist of two flat conducting plates separated by a small gap. The potential difference, or voltage, between the plates is proportional to the difference in the amount of the charge on the plates. Capacitance is the amount of charge that can be stored per unit of voltage. The unit for measuring capacitance is the farad (F), named for physicist Michael Faraday, and is defined as the capacity to store one coulomb of charge with an applied potential of one volt. One coulomb (C) is the amount of charge transferred by a current of one ampere in a second.

To maximise efficiency, capacitor plates are stacked in layers or wound in coils with a very small air gap between them. Dielectric materials – insulating materials that partially block the electric field between the plates – are often used within the air gap. This allows the plates to store more charge without arcing and shorting out.

Capacitors are often found in active electronic circuits that use oscillating electric signals, such as those in radios and audio equipment. They can charge and discharge nearly instantaneously, which allows them to be used to produce or filter certain frequencies in circuits. An oscillating signal can charge one plate of the capacitor while the other plate discharges, and then, when the current is reversed, it will charge the other plate while the first plate discharges.

In general, higher frequencies can pass through the capacitor, while lower frequencies are blocked. The size of the capacitor determines the cutoff frequency for which signals are blocked or allowed to pass. Combinations of capacitors can be used to filter selected frequencies within a specified range.

Stronger supercapacitors are manufactured using nanotechnology to create super-thin layers of materials, such as graphene, that achieve capacities that are 10 to 100 times that of conventional capacitors of the same size. However, they have much slower response times than conventional dielectric capacitors, so they can't be used in active circuits.

These devices store charge

Creating a capacitor
How does this device hold a small reserve of energy?

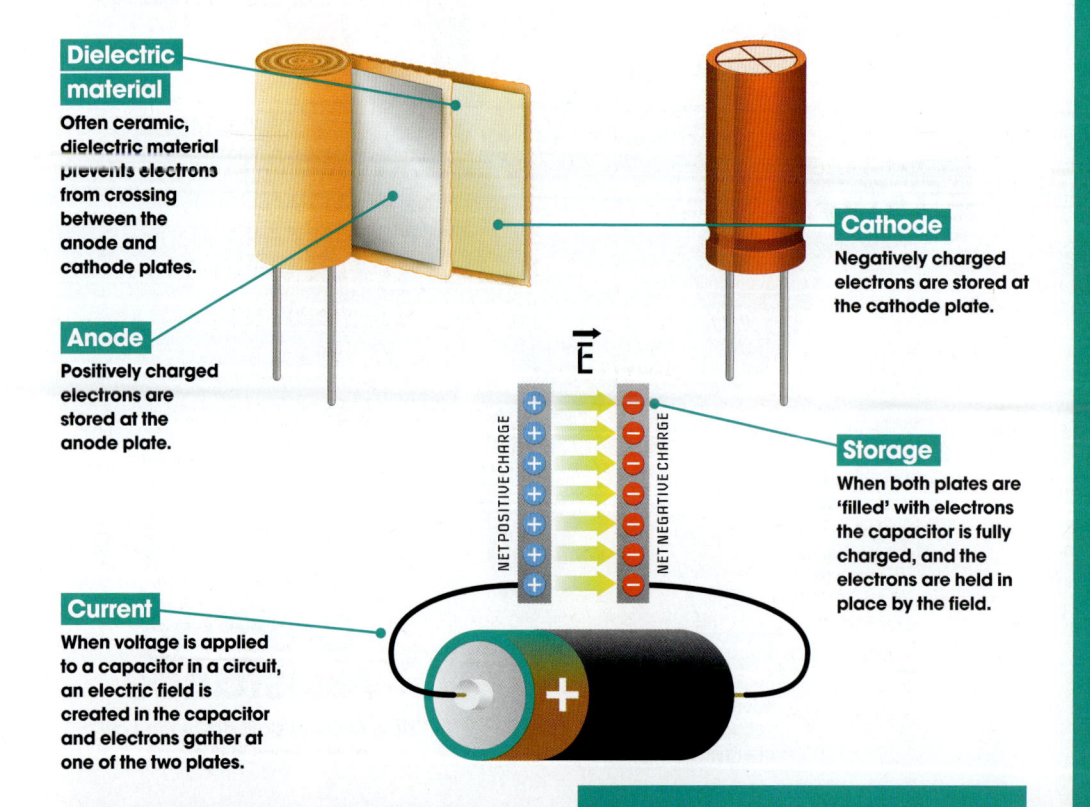

Dielectric material
Often ceramic, dielectric material prevents electrons from crossing between the anode and cathode plates.

Anode
Positively charged electrons are stored at the anode plate.

Current
When voltage is applied to a capacitor in a circuit, an electric field is created in the capacitor and electrons gather at one of the two plates.

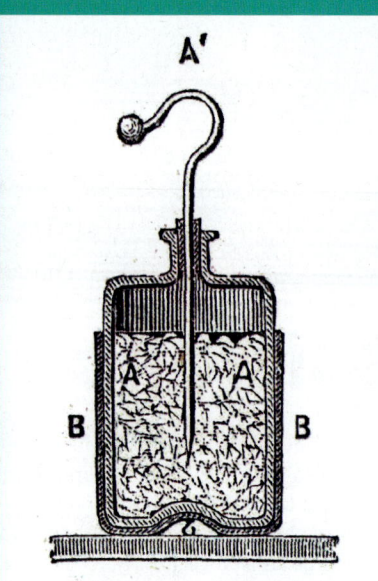

Cathode
Negatively charged electrons are stored at the cathode plate.

Storage
When both plates are 'filled' with electrons the capacitor is fully charged, and the electrons are held in place by the field.

FIG. 142. — Bouteille de Leyde. — A, armature intérieure; B, armature extérieure.

A drawing of a Leyden jar, a piece of apparatus used to store electric charge, invented in 1745

Inside the Leyden jar

The earliest example of a capacitor is the Leyden jar. This device was invented to store a static electric charge in conducting foil that lined the inside and outside of a glass jar. The jar was created by Ewald von Kleist and Pieter van Musschenbroek, both working independently during the early 1740s. Musschenbroek was a teacher at the University of Leiden in The Netherlands, and so named this device the Leyden jar. The glass jar housed two sheets of foil that acted as conductors – one on the outside of the jar and the other lining it. A metal chain connected to an iron rod extended through a wooden lid with a ball at the end. When a charge was applied to the conductors, the electrons were temporarily trapped and stored.

EVERYDAY PHYSICS

ELECTRICITY BASICS

Inductance

An inductor is an electronic component consisting of a coil of wire with an electric current running through it, creating a magnetic field. The unit for inductance is the henry (H), named after Joseph Henry, an American physicist who discovered inductance independently at about the same time as English physicist Michael Faraday. One henry is the amount of inductance that's required to induce one volt of electromotive force – the electrical pressure from an energy source – when the current is changing at one ampere per second.

One important application of inductors in active circuits is that they tend to block high-frequency signals while letting lower frequency oscillations pass. This is the opposite function of capacitors. Combining the two components in a circuit can selectively filter or generate oscillations of almost any desired frequency. With the advent of integrated circuits, such as microchips, inductors are becoming less common, because 3D coils are extremely difficult to fabricate in 2D-printed circuits. For this reason, microcircuits are designed without inductors, instead using capacitors to achieve essentially the same results.

Coils create a magnetic field

A shocking discovery

American author, scientist and diplomat Benjamin Franklin is often credited as the discoverer of electricity after flying a kite during a lightning storm. His iconic experiment involved him attaching a wire to the top of a kite – which was a precursor to the lightning rod – tied to a hemp string, which would have been made sodden by the rain, and a silk string for Franklin to hold. Although it's a misconception that the lightning hit the kite directly, the flying experiment produced a spark felt by Franklin, and the fibres of the hemp string stood erect, much like the hairs on your arm when you rub it with a balloon.

Franklin demonstrated the connection between lightning and electricity, but he was not the first. Thousands of years ago, the ancient Greeks carried out static electrical experiments using fur and amber. And in Iraq around 2,000 years ago, jars called Baghdad Batteries used sheets of copper and an iron rod to attempt to create light.

Benjamin Franklin flying a kite during a thunderstorm in 1752

Inside an inductor
How a magnetic field can support a circuit

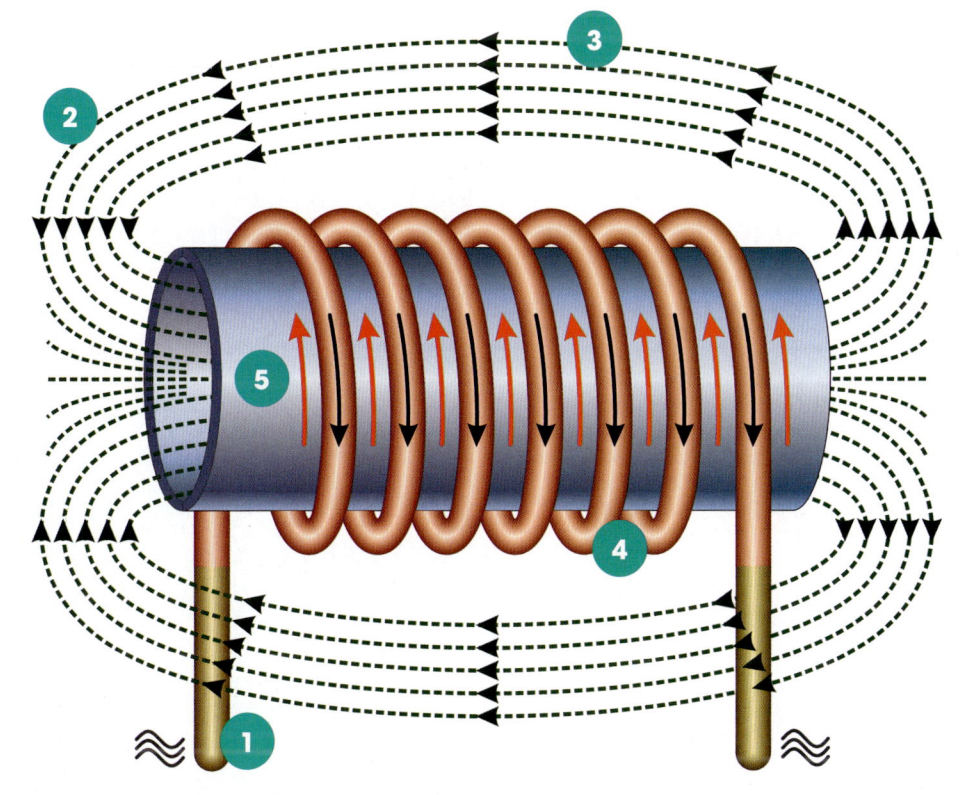

"They block high-frequency signals while letting lower frequency oscillations pass"

1 Wire
An electrical current is passed through a copper wire, and a magnetic field, or flux, is generated around it.

2 Magnetic field direction
The magnetic field follows the direction that the current is flowing. When the current changes direction, so does the magnetic field.

3 Energy outage
When the current is turned off, the energy stored in the magnetic field will continue to supply power, but only for a short period of time.

4 Power up
The greater the amount of current passed through the coil, the larger the magnetic field produced.

5 Core
Copper wires are commonly wrapped around a non-conductive material, such as plastic.

UNDERSTANDING PHYSICS

The science behind your swing

Find out why tennis is such a difficult sport to perfect

Many sports scientists would argue that tennis is the toughest sport out there. It requires a mixture of speed, strength, endurance and mental fortitude, not to mention an abundance of talent if you plan on reaching the game's highest level.

Every professional strives to make perfect contact with the ball on each shot they hit. Scientists have proven that this is much harder to achieve than you might imagine, by calculating how far you would miss your target if you changed the racket angle by only one degree. It turns out that this tiny alteration would cause you to miss your target by an enormous 41 centimetres (16 inches). Further to this, they calculated that to make perfect contact you have a small window of only 0.6 thousandths of a second, presuming that the racket angle changes throughout the swing. This shows just how hard tennis is to master and why even the most seasoned professionals still need to practice for hours each day.

Both the player and the court can alter the shot

It's not just the players that affect the way a ball bounces and moves through the court, the court itself is also important. Tennis is played on a variety of contrasting surfaces, all of which affect a ball's trajectory and speed. A clay court will steal part of a shot's momentum due to the friction between the loose clay and the ball. This will slow down a 107.8-kilometre (67-mile) per hour shot by 43 per cent to only 61 kilometres (38 miles) per hour, giving the opponent extra time to return the shot. This differs to the grass courts of Wimbledon, which would maintain a speed of roughly 72.4 kilometres (45 miles) per hour for the same shot. Memories of previous Wimbledon tournament highlights may now be fading for many, but it will undoubtedly have inspired many of us to perfect our swing in the hope of grass court glory.

Secrets of spin

Modern professionals tend to be looking for more control rather than power in their game, and the key is to use topspin. This is achieved by the speed at which the racket is swung, the angle it connects with the ball, and the type of strings that are used. The polyester strings on modern rackets impart massive amounts of grip on the ball, acting almost like suction cups. This extra control enables players to generate large amounts of topspin: by brushing up the back of the ball with the strings causes the ball to rotate forward, which in turn creates an area of high pressure above the ball and low pressure below, so the ball dips sharply once the spin has taken effect. This means that players can hit the ball harder and still get it to land inside the lines. It will also give their opponents more problems as the ball bounces higher, making it even harder to return.

Rafael Nadal is widely considered to be the king when it comes to generating topspin, having had some of his forehands recorded at an incredible 4,900 revolutions per minute (rpm). However, when it comes to backhand spin, Roger Federer is capable of generating a whopping 5,300 rpm!

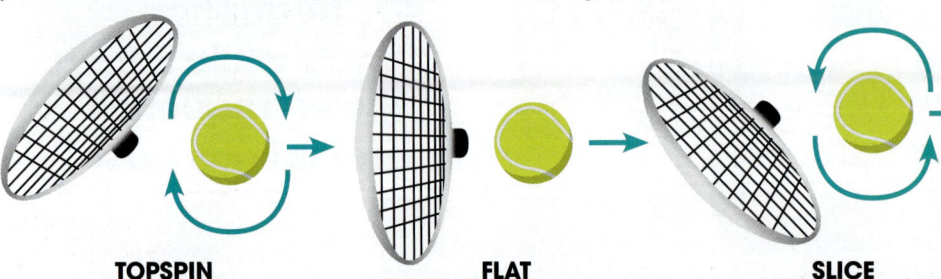

TOPSPIN
This is the most commonly used type of spin, particularly prevalent in forehands, backhands and second serves. It causes the ball to jump up off the court, so that it bounces to an awkward height to return.

FLAT
Today, most professionals hit a flat first serve as this will enable them to hit with the most power. This is very effective on fast courts, as the ball will skid through quickly and take time away from the opponent.

SLICE
The main advantage of the slice technique is that it keeps the ball low. It works well on fast, low bouncing courts, forcing the opponent to bend down and return the shot from an uncomfortable angle.

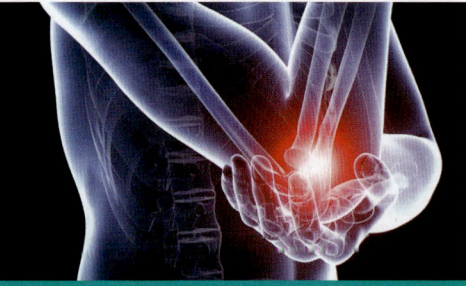

Approximately five in every 1,000 people in the UK visit their GP about tennis elbow

Tennis elbow

Technically referred to as lateral epicondylitis, tennis elbow is an injury that plagues many tennis players, leaving them unable to compete. It causes pain around the outside of the elbow, occurring when the forearm's muscles and tendons are strained as a result of repetitive or strenuous activity. Minute tears can form in the muscles surrounding the joint, causing it to become inflamed. This can result in a range of symptoms, varying from only mild discomfort when the elbow is in use, to severe pain even when resting the elbow. There is unfortunately no quick fix for this condition.

Tennis elbow can last for months or even longer, due to the slow speed at which tendons heal. The best treatment is to rest the elbow as much as possible and in 90 per cent of cases it will heal. Doctors may recommed physiotherapy or surgery for severe cases, however.

EVERYDAY PHYSICS

THE SCIENCE BEHIND YOUR SWING

Anatomy of the serve
Learn the biomechanics behind the most important shot in tennis

Wrist snap
Immediately before contact is made, the wrist bends back away from the ball and then quickly 'snaps' forward, throwing the racket into the ball. This small movement provides 30 per cent of the racket-head speed, which is why a strong wrist is essential for all tennis players.

Shoulder strength
The upper arm and shoulder provide only ten per cent of the racket-head speed, but are still vitally important. They help form what's known as the tick position which is the point just before the server swings the racket at the ball.

Racket-head speed
All of a tennis player's muscles work together to increase their racket-head speed, which is the speed at which their racket is propelled towards the ball.

Forearm power
The forearm contributes 40 per cent of the total racket-head speed, the most out of any single body part. Experts recommend that the arm be fully extended at the point of contact, as this increases the overall length of the lever, providing the greatest possible speed.

Coiled trunk
The trunk uncoils after the hips, continuing the uncoiling process of the kinetic chain. A fast, powerful trunk rotation will translate into a more effective serve. Depending on the server's style, a large hip rotation can add considerable speed to their serve.

Knee bend
The legs and trunk provide 20 per cent of a serve's overall power. Some players use an exaggerated knee bend to help them spring up into the serve, which helps them make contact with the serve as high as possible. This provides them with a larger area to hit the serve into.

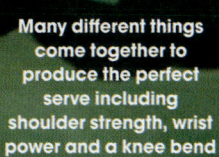

Many different things come together to produce the perfect serve including shoulder strength, wrist power and a knee bend

1 Stepping up to the line
The server places his leading foot close to the line, making sure not to touch it before making contact with the ball, as this would be considered a foot fault. His feet are spaced quite far apart, providing a wide base to push up from once the motion starts.

2 The toss
For fast serves, players throw their ball toss up to 60cm (24 inches) inside the baseline. This enables them to lean into the serve and generate power through weight transfer, as they uncoil upwards and forwards simultaneously.

3 Tick position
Nearly all players adopt this position immediately before they contact the ball. Many players leave their non-dominant arm in a raised position to balance out the action of the racket arm, and also keep their torso raised which helps reduce the chance of serving the ball into the net.

4 The follow through
The forward motion generated by the serve will often cause the server's body to be thrown forwards so that they land inside the baseline. Hip rotation is an excellent source of power and we can see in this image that the server's body has turned to face the court after impact. His right leg extends backwards to provide balance and to enable him to quickly prepare for the next shot.

UNDERSTANDING PHYSICS

The big theory

36 A beginner's guide to time travel
Why there's nothing in science that says time travel is impossible

44 Newton's laws of motion
Three simple laws explain the effect of forces on the world around us

45 The general theory of relativity
Get to grips with Einstein's theory of the universe

46 What is string theory?
This strange idea could explain how the entire universe works

48 Matter vs antimatter
From space drives to bananas – everything you need to know about the mysteries of antimatter

54 Testing Hawking's theories
Which of Stephen Hawking's explanations turned out to be right?

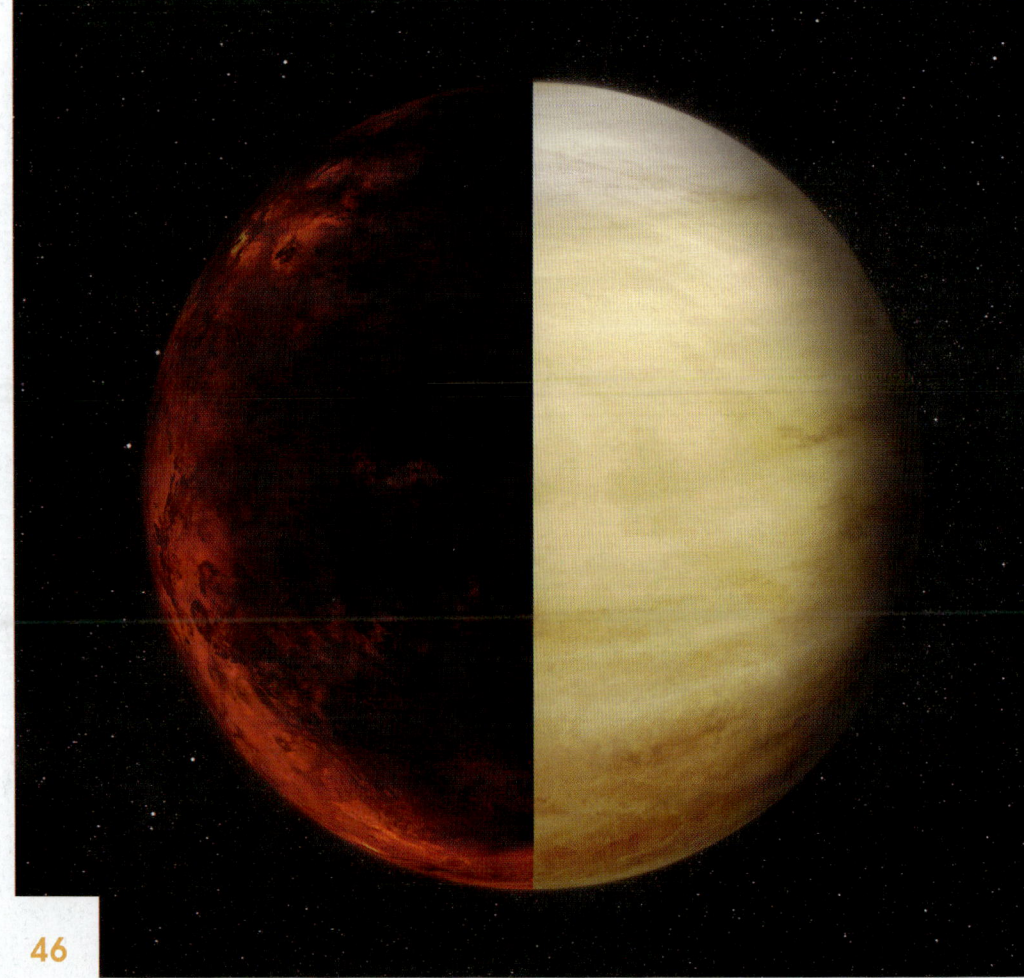

THE BIG THEORY

45

48

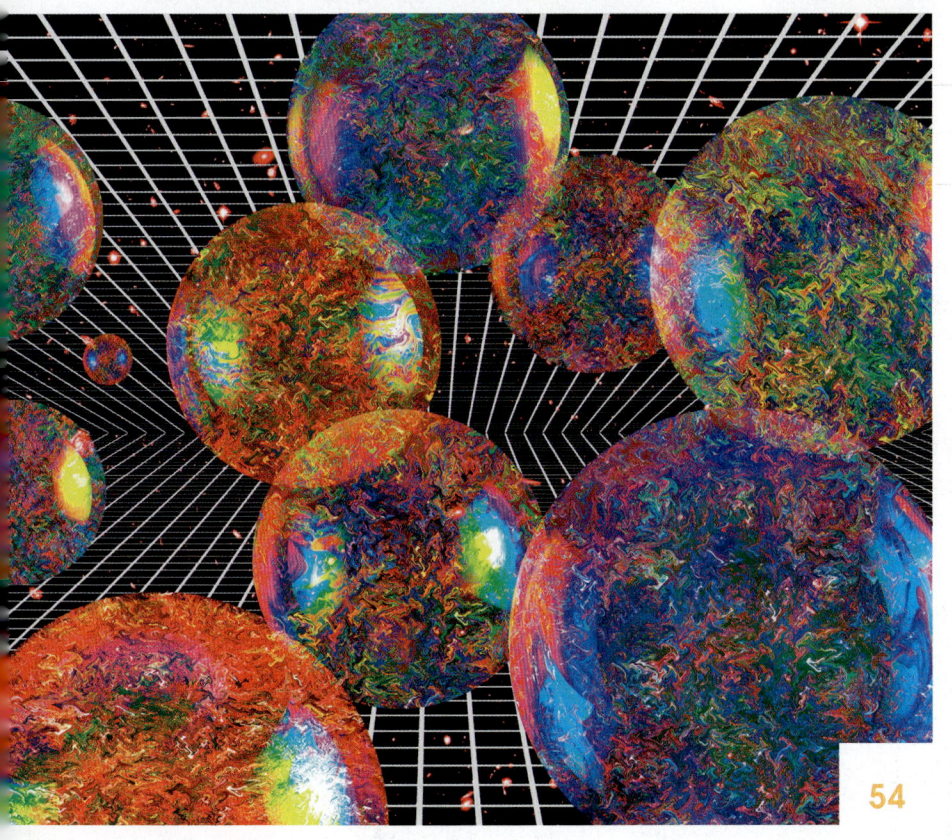

54

44

UNDERSTANDING PHYSICS

A beginner's guide to time travel

Learn exactly how Einstein's theory of relativity works, and discover how there's nothing in science that says time travel is impossible

Everyone can travel in time. You do it whether you want to or not, at a steady rate of one second per second. You may think there's no similarity to travelling in one of the three spatial dimensions at say, one metre per second. But according to Einstein's theory of relativity, we live in a four-dimensional continuum – spacetime – in which space and time are interchangeable.

Einstein found that the faster you move through space, the slower you move through time – you age more slowly, in other words. One of the key ideas in relativity is that nothing can travel faster than the speed of light – about 300,000 kilometres per second (or one light year per year). But you can get very close to it. If a spaceship was flying at 99 per cent of the speed of light, you'd see it travel a light year of distance in just over a year of time.

That's obvious enough, but now comes the weird part. For astronauts onboard, the journey would take a mere seven weeks. It's a consequence of relativity called time dilation, and in effect it means the astronauts have jumped about ten months into the future.

THE BIG THEORY

A BEGINNER'S GUIDE TO TIME TRAVEL

UNDERSTANDING PHYSICS

Travelling at high speed isn't the only way to produce time dilation. Einstein showed that gravitational fields produce a similar effect – even the relatively weak field here on the surface of the Earth. We don't notice it, because we spend all our lives here, but 20,000 kilometres higher up gravity is measurably weaker – and time passes more quickly, by about 45 microseconds per day. That's more significant than you might think, because it's the altitude at which GPS satellites orbit the Earth, and their clocks need to be precisely synchronised with ground-based ones for the system to work properly. The satellites have to compensate for time dilation effects due both to their higher altitude and their faster speed. So whenever you use the GPS feature on your smartphone or your car's satnav, there's a tiny element of time travel involved. You and the satellites are travelling into the future at very slightly different rates.

But for more dramatic effects, we need to look at much stronger gravitational fields, such as those around black holes, which can distort spacetime so much that it folds back on itself. The result is a so-called wormhole, a concept that's familiar from sci-fi movies but actually

Albert Einstein, whose name has become virtually synonymous with relativity

Real time machines are unlikely to be as simple as this 1960 Hollywood portrayal

A brief history of time travel

1895
H.G. Wells's novel The Time Machine popularises the idea of time as the fourth dimension, through which it might be possible to travel by analogy with the three spatial dimensions.

1905
Einstein's groundbreaking paper on the theory of relativity introduces the idea of time dilation – the first hint that, in real physics as well as sci-fi, time might be interchangeable with space.

1927
The physicist Arthur Eddington first introduces the concept of the 'arrow of time', and its relation to entropy, in his book titled The Nature Of The Physical World.

1935
Together with Nathan Rosen, Einstein shows that under certain circumstances it's possible to have a shortcut between two different points in spacetime – a wormhole – even between past and future.

THE BIG THEORY

A BEGINNER'S GUIDE TO TIME TRAVEL

"Everything not forbidden is compulsory"

originates in Einstein's theory of relativity. In effect, a wormhole is a shortcut from one point in spacetime to another. You enter one black hole and emerge from another one somewhere else. Unfortunately, it's nothing like as practical a means of transport as Hollywood makes it look, because the black hole's gravity would tear you to pieces as you approached it, but it really is possible in theory. And because we're talking about spacetime, not just space, the wormhole's exit could well be at an earlier time than its entrance, so you'd end up in the past rather than the future.

Trajectories in spacetime that loop back into the past are given the technical name 'closed timelike curves'. If you search through serious academic journals, you'll find plenty of references to them – far more than you'll find to 'time travel'. But in effect, that's exactly what closed timelike curves are all about.

There's another way to produce a closed timelike curve that doesn't involve anything quite so exotic as a black hole or wormhole: a simple a rotating cylinder made of super-dense material. Called a Tipler cylinder, it's the closest that real-world physics can get to an actual, genuine time machine. But it's not something that's ever likely to be built in practice, so like a wormhole it's more of an academic curiosity than a viable engineering design.

Yet as far-fetched as these things are in practical terms, there's no fundamental scientific reason – that we currently know of – that says they're impossible. That's a thought-provoking situation, because as the physicist Michio Kaku is fond of saying – "everything not forbidden is compulsory". He doesn't mean time travel has to happen everywhere all the time, but that the universe is so vast it ought to happen somewhere at least occasionally. Maybe some super-advanced civilisation in another galaxy knows how to build a working time machine, or perhaps closed timelike curves can even occur naturally under certain rare conditions.

This raises problems of a different kind – not in science or engineering, but in basic logic. If time travel is allowed by the laws of physics, then it's possible to envisage a whole range of paradoxical scenarios. Some of these appear so illogical that it's difficult to imagine that they could ever occur. But if they can't, what's stopping them?

Thoughts like these prompted Stephen Hawking, who was always sceptical about the idea of time travel into the past, to come up with his 'chronology protection conjecture': the notion that some as-yet-unknown law of physics prevents closed timelike curves from happening. But it's only an educated guess, and until it's supported by hard evidence, there's only one conclusion we can come to: time travel is possible.

A party for time travellers

Physicist Stephen Hawking was sceptical about the feasibility of time travel into the past. This wasn't because he'd disproved it, but he was bothered by the logical paradoxes it created. In his 'chronology protection conjecture', he surmised that physicists would eventually discover a flaw in the theory of closed timelike curves that made them impossible.

In 2009 he came up with an amusing way to test this conjecture. Hawking held a champagne party (shown in his Discovery Channel programme) but he only advertised it after it had happened. His reasoning was that, if time machines eventually become practical, someone in the future might read about the party and travel back to attend it. But no one did – Hawking sat through the whole evening on his own. This doesn't prove time travel is impossible, but it does suggest that it never becomes a commonplace occurrence here on Earth.

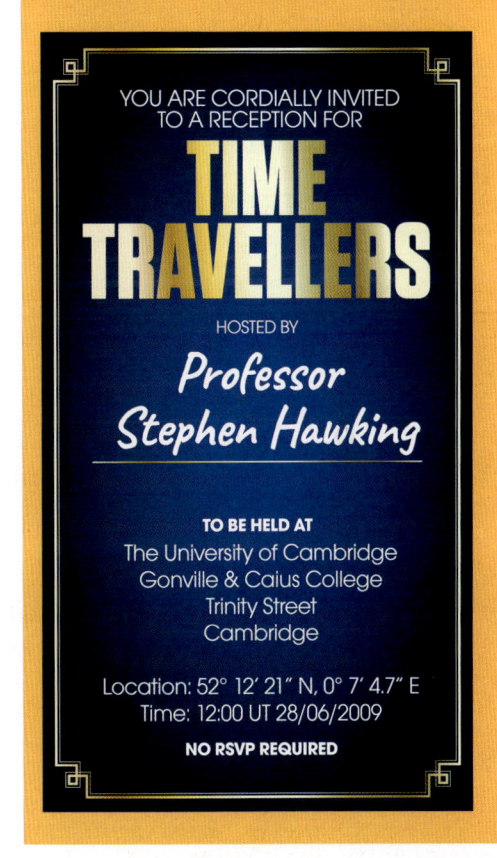

YOU ARE CORDIALLY INVITED TO A RECEPTION FOR

TIME TRAVELLERS

HOSTED BY

Professor Stephen Hawking

TO BE HELD AT
The University of Cambridge
Gonville & Caius College
Trinity Street
Cambridge

Location: 52° 12' 21" N, 0° 7' 4.7" E
Time: 12:00 UT 28/06/2009

NO RSVP REQUIRED

1941
Two American experimenters, Herbert Ives and G.R. Stilwell, confirm the reality of time dilation by observing fast-moving particles inside a TV-style cathode ray tube.

1974
Physicist Frank Tipler designs the first real time machine (on paper at least). According to the design, a Tipler cylinder would use a string of rotating neutron stars to produce a closed timelike curve.

1992
Stephen Hawking suggests that there might be an undiscovered law of nature that prevents closed timelike curves, therefore preventing time travel into the past from occurring.

2009
Stephen Hawking holds a party for time travellers, which is widely publicised – but only after it has taken place. Unfortunately no time travellers turn up to the party.

UNDERSTANDING PHYSICS

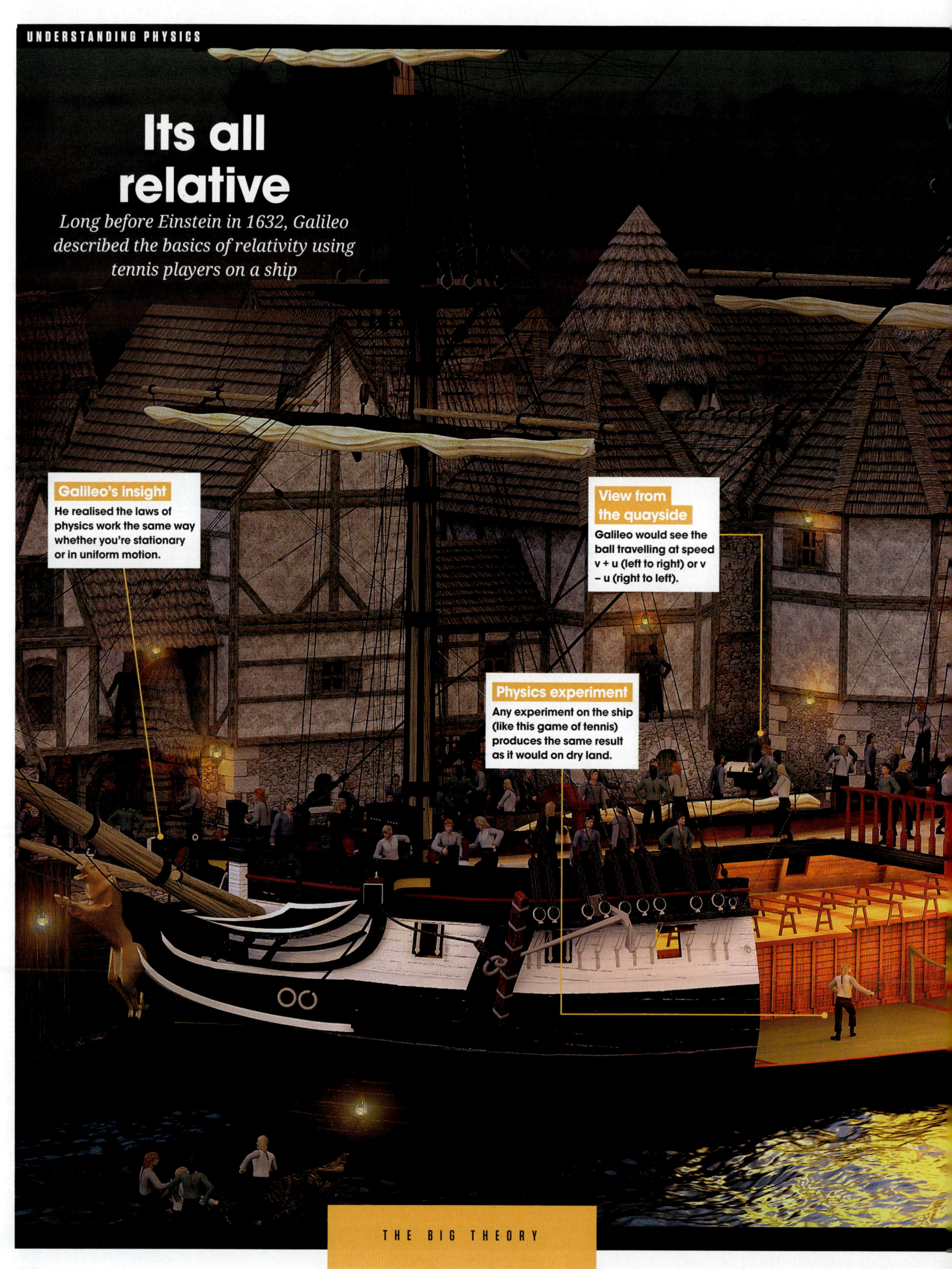

Its all relative

Long before Einstein in 1632, Galileo described the basics of relativity using tennis players on a ship

Galileo's insight
He realised the laws of physics work the same way whether you're stationary or in uniform motion.

View from the quayside
Galileo would see the ball travelling at speed v + u (left to right) or v − u (right to left).

Physics experiment
Any experiment on the ship (like this game of tennis) produces the same result as it would on dry land.

THE BIG THEORY

A BEGINNER'S GUIDE TO TIME TRAVEL

GPS satellites can help you navigate accurately, but only after compensating for relativistic time dilation

Ship speed v
As long as this stays constant (and the sea is calm) the people inside are unaware of the motion.

Einstein's twist
Light isn't like a tennis ball – its speed doesn't add and subtract, but stays the same for all observers.

Ball speed u
The players see the ball travelling at speed u, exactly the same as on a stationary tennis court.

The arrow of time

One of the distinctive things about time is that it has a direction – from past to future. A cup of hot coffee left at room temperature always cools down, it never heats up. Your mobile phone loses battery charge when you use it – it never gains charge. These are examples of entropy, essentially a measure of the amount of 'useless' as opposed to 'useful' energy. The entropy of a closed system always increases, and it's the key factor determining the arrow of time.

It turns out that entropy is the only thing that makes a distinction between past and future. In other branches of physics, like relativity or quantum theory, time doesn't have a preferred direction. No one knows where time's arrow comes from. It may be that it only applies to large, complex systems, in which case subatomic particles may not experience the arrow of time.

Time has a clear flow from past to future, but why is a mystery

Time travel paradox

If it's possible to travel back into the past – even theoretically – it raises a number of brain-twisting paradoxes that even scientists and philosophers find extremely perplexing.

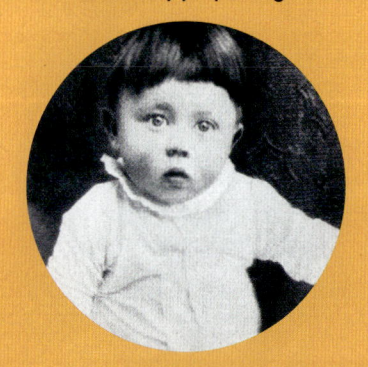

Killing Hitler

Hitler was one of the most evil people in history, causing untold death and misery. A time traveller might decide to go back and kill him in his infancy. If they succeeded, future history books wouldn't even mention Hitler – so what motivation would the time traveller have for going back in time and killing him?

Killing your grandfather

Instead of killing a young Hitler, you might, by accident, kill one of your own ancestors when they were very young. But then you would never be born, so you couldn't travel back in time to kill them, so you would be born after all, and so on...

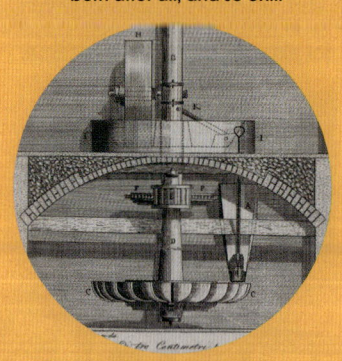

A closed loop

Suppose the plans for a time machine suddenly appear from thin air on your desk. You spend a few days building it, then use it to send the plans back to your earlier self. But where did those plans originate? Nowhere – they're just looping round and round in time.

The relativity of time and space

The fact that all observers measure the same speed of light has some odd consequences

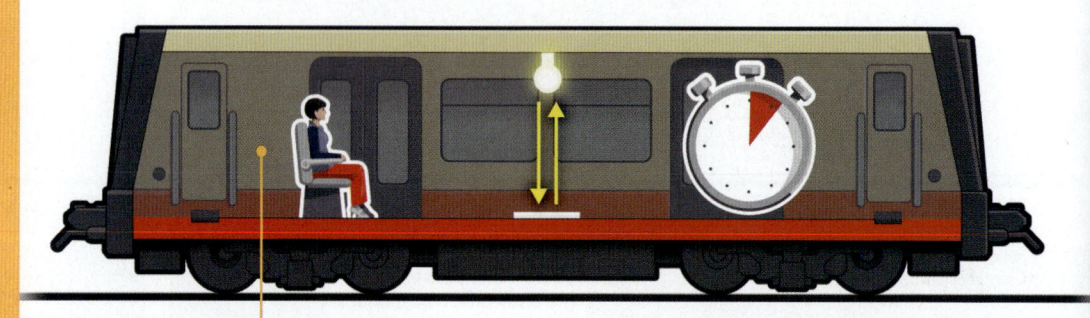

1 Experiment inside a moving train
Alice measures the time a beam of light takes to travel from the ceiling of the carriage to a mirror on the floor, and then back to the ceiling.

4 Another experiment
Alice uses a laser rangefinder to measure the carriage length. This is the time the beam takes to travel to the end and back, divided by the speed of light.

Tipler cylinder

In 1974, the eminently respectable journal Physical Review published a paper by Frank Tipler that contained the first scientifically feasible design for a time machine. Sadly, although it doesn't break any laws of physics, Tipler's machine poses so many engineering challenges it would be impossible to build one in practice.
   The idea is to generate a closed timelike curve using a long, rapidly rotating cylinder of extremely dense material, of the kind found in neutron stars. These aren't quite as extreme as black holes, but they're still not easy things to manipulate. And to make a cylinder of the necessary length, at least ten of them would need to be brought together and lined up. This means that, just like wormholes, Tipler cylinders fall into the 'possible in theory, but impossible in practice' category.

THE BIG THEORY

A BEGINNER'S GUIDE TO TIME TRAVEL

2 The view from outside
Outside the train, Bob sees the light travelling a greater distance – but at the same speed. For him, the trip from the ceiling to the floor and back takes longer.

3 Time is relative
Bob, as a stationary observer, sees events inside the moving train happening more slowly than Alice perceives them from the interior of the train.

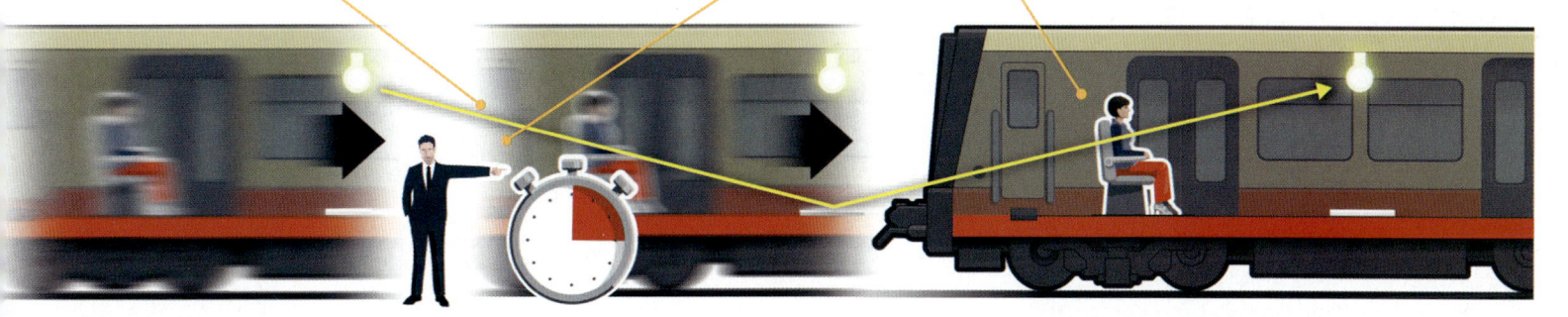

5 Bob's view
Seen from outside, the end of the carriage is moving towards the laser, so it takes less time to reach it. In other words, the carriage must be shorter.

6 Length is relative
Alice, moving with the train, sees it at its normal length. But to Bob, who remains stationary, it appears to have contracted along the direction of motion.

> "The idea is to generate a closed timelike curve using a long, rapidly rotating cylinder of extremely dense material, of the kind found in neutron stars"

An artist's impression of a pair of neutron stars – a Tipler cylinder requires at least ten

UNDERSTANDING PHYSICS

Newton's Laws of Motion

Three simple laws explain the effect of forces on the universe around us

Background

Isaac Newton's famous Laws of Motion explain what happens to objects when forces are applied. A force is a push or a pull, like gravity, friction or magnetism. They can't be seen directly, but their effects can be measured; they can change the speed, shape or direction of movement of an object, and they are responsible for pressure and weight. Newton's three laws describe what happens when forces are balanced or unbalanced, and explain the idea of equal and opposite forces.

In brief

Newton's First Law explains what happens if the forces acting on an object are balanced. If an object is not moving, it won't start moving. And, if an object is already moving, it won't stop. This tendency is known as inertia.

Newton's Second Law describes what happens if the forces acting on an object are unbalanced. If more force is applied in one direction, the object will accelerate. The more unbalanced the forces, the faster the object will accelerate. The more massive the object, the more force that is needed to make it move.

Newton's Third Law explains that for every action there is an equal and opposite reaction. Forces come in pairs; if one object exerts a force on another, the first object will exert an equal force in return. A simple example is the recoil of a gun; as the bullet flies forwards, the gun kicks back.

Newton's laws first appeared in his masterpiece, Principia, in 1687, and he developed them to explain why the orbits of the planets are ellipses, not circles.

Summary

Newton's First Law describes what happens when forces are balanced. His Second Law describes what happens when they are unbalanced. The Third Law explains forces acting in equal and opposite pairs.

Newton's Laws in action

The Laws of Motion govern the movement of everything around us

At rest

First Law
The forces acting on the stationary rocket are balanced. The downward pull of gravity is matched by the upward push of the ground.

Normal force
The Earth exerts an upward force on the rocket.

Second Law
As the engines fire, the force of the thrust is greater than the force of gravity. They become unbalanced and the rocket accelerates.

Air resistance
A frictional force acts on the rocket as it moves through the air.

Applied force
The exhaust from the engine applies a force beneath the rocket.

Gravity
Objects with mass are attracted to one another by the force of gravity.

Third Law
The force that pushes exhaust gas out of the rocket is matched by an equal and opposite force – thrust.

The man behind the Laws

Sir Isaac Newton was a mathematician, physicist and astronomer, born on Christmas day in 1642 (according to the old julian calendar). He described the mechanics of the universe with maths and equations in his book, The Philosophiae Naturalis Principia Mathematica (commonly known as PRINCIPIA). He explained the concept of gravity, and showed that everything in the universe is governed by the same physical laws. He also worked on colour theory, optics and calculus, and his ideas are still in use over 300 years later.

He was one of the greatest scientists ever to have lived, but his achievements didn't stop there. He built the first practical reflecting telescope and was elected as a Member of Parliament. He even became Master of the Royal Mint, in charge of the production of all of Britain's currency from 1699 until his death in 1727.

THE BIG THEORY

THE GENERAL THEORY OF RELATIVITY

The general theory of relativity

Get to grips with Einstein's theory of the universe

Background

In 1905, Albert Einstein published his theory of special relativity, explaining that the speed of light in a vacuum is constant and so are the laws of physics when they are observed while not accelerating. He proved that everything moves relative to everything else, but it only applied to special cases; it did not apply to observers who were speeding up or slowing down. Einstein set about extending his theory so that it could apply to everything in the universe, forming a theory of general relativity.

In brief

According to Isaac Newton's first law of motion, objects do not accelerate unless an external force acts upon them. However, Einstein realised that when you are in freefall, you feel weightless, so you feel no force even though you're accelerating towards the ground. He determined that what we experience as gravity must be the result of massive objects curving space-time itself. Any objects moving through this warped space-time follow as short a path as possible, which is a curve. This helped to prove that Earth's orbit was not determined by gravity pulling it towards the Sun, as had been previously thought, but was rather the result of curved space-time forcing our planet along the shortest possible route around its host star.

Summary

The theory of general relativity proves that gravity is caused by the curvature of space-time and does not pull objects, but instead forces them along the shortest possible path.

Bending space-time
Explaining motion and the path of light in space

Curved space-time
Space-time can be visualised using the analogy of a flat sheet that bends under the influence of massive objects such as the Sun, like a bowling ball on a trampoline.

Star position
The gravity of massive objects also bends light, causing the apparent position of stars to shift when they are viewed from Earth.

Planetary orbit
Objects in space move along the straightest possible route, which in space-time is the curved path around a massive object.

Black holes
Extremely massive objects bend space-time so much that nothing – not even light – can escape.

How general relativity changed the world

- Einstein had solved the mystery of where gravity comes from – the curving of space-time.
- It was discovered that the curvature of space-time around extremely dense objects is infinite, forming a hole in the fabric of space-time, known as a black hole.
- Using general relativity, Einstein proved that gravity bends the path of light and gives stars a false position in the sky when seen from Earth.
- The equations of general relativity helped reveal that the universe is expanding, leading to the development of the big bang theory.

Albert Einstein
1879-1955

Einstein considered his general theory to be the culmination of his life's research. After it was published in 1915, he became world famous almost overnight and in 1921, was awarded the Nobel Prize for Physics. He published more than 300 scientific papers in his lifetime, changing the world's view on space, time and matter.

UNDERSTANDING PHYSICS

What is string theory?

Could this strange idea explain how the entire universe works?

String theory, or more accurately superstring theory, is the idea that elementary particles like electrons are made up of vibrating strings of energy. While we once thought elementary particles were the smallest things possible, string theory suggests that there is something even smaller that makes up the universe around us.

The theory is also known as M-theory, and it's intended as a way to explain some of the limitations of the long-established Standard Model of physics. This is the model we use to explain how everything in the universe works, but it breaks down when we look at things at a quantum level – the level at which weird effects take place, like particles appearing in two positions at once (called superposition) or being able to share information over great distances (called entanglement).

As the strings move through time they vibrate in one dimension in different patterns, or 'modes'. Each one of these can make the string appear like an electron, a photon, and so on. At bigger scales these strings simply look like particles to us. Some suggest string theory could be the much-sought-after, unifying 'theory of everything'. It can also explain how two particles of gravity, known as gravitons, can interact on large scales, when other theories cannot.

Not everyone believes in string theory, however. One notable problem is that it requires space-time to have at least ten dimensions, six more than our standard four: three for space and one for time. It's suggested that these extra six dimensions are so compact that we don't even know they're there. Plus, it's very hard to prove. We can't really measure these strings, so how will we ever know that they exist? Nonetheless, string theory continues to be debated by the scientific community.

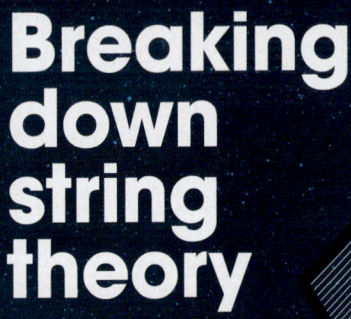

Breaking down string theory

Get ready – things are about to get complicated

The graviton
String theory could help explain how gravity is able to take place at vast distances using particles called gravitons.

Particles
As the vibrations move, they trace out 'tubes' that form things like electrons and photons.

Proving string theory
It's difficult or perhaps impossible for us to test string theory, making its existence controversial.

Quantum world
Our current models cannot explain both the large-scale universe and small-scale quantum world. But string theory can.

THE BIG THEORY

WHAT IS STRING THEORY?

Superstring
The 'super' in superstring theory refers to supersymmetry, a realm of physics beyond the Standard Model.

Theory of everything
String theory could be a theory of everything, helping us understand how the entire universe works.

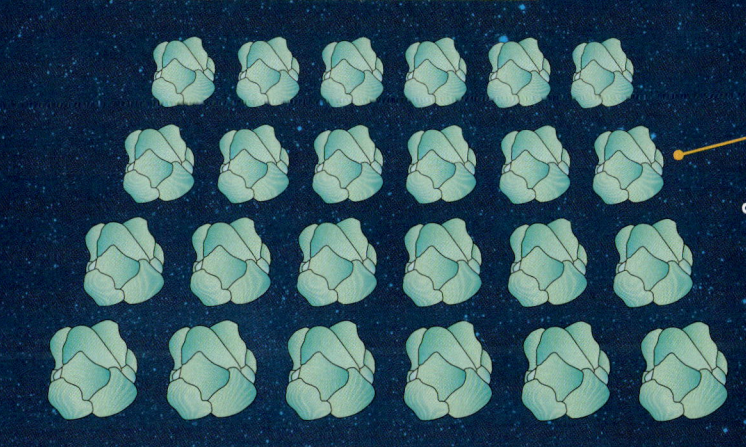

Ten dimensions
We generally regard the universe as having four dimensions. String theory would require six additional dimensions to work.

Bad vibrations
But under string theory, these vibrations are so small that we can never actually see them.

Good vibrations
According to string theory, all particles can be broken down into vibrating strings of energy.

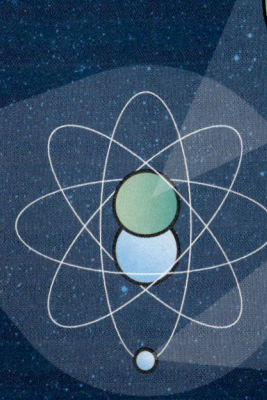

"String theory could be the 'theory of everything'"

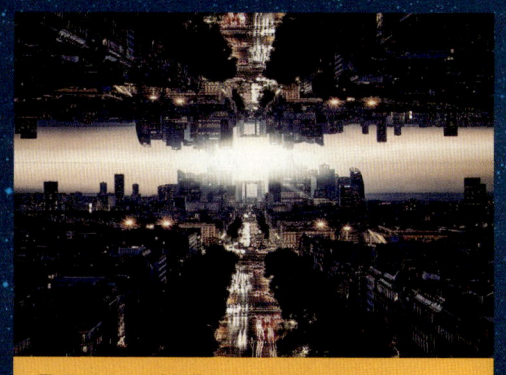

The many-worlds theory predicts there are near-infinite parallel universes

Parallel universes
String theory isn't the only controversial theory on the block. Another is the many-worlds interpretation, which suggests that the universe has an almost infinite number of parallel universes. First put forward in 1957, it suggests that everything is quantum, meaning that things can occur in multiple ways simultaneously. If you apply this at both a small and a large scale, it means that all possibilities in any given scenario should occur, with each giving rise to a universe that is equally real. Under some versions of the theory it could actually be possible to see the effects of parallel universes on each other.

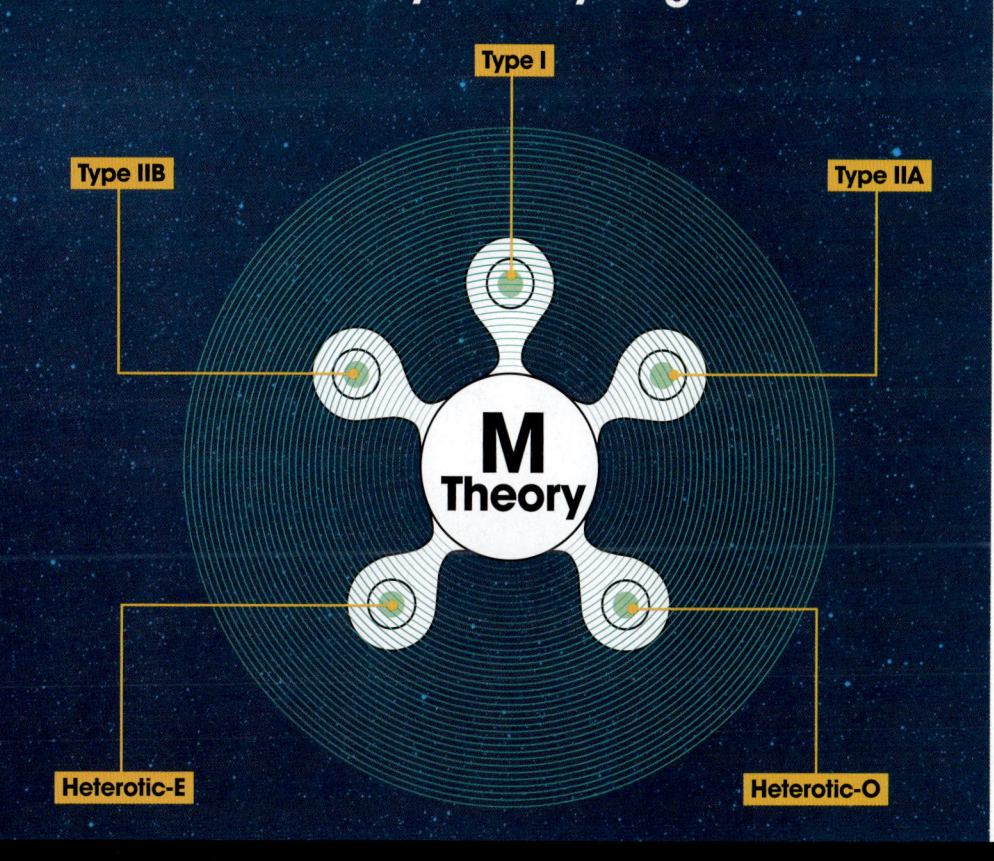

M Theory
- Type I
- Type IIA
- Type IIB
- Heterotic-E
- Heterotic-O

047

UNDERSTANDING PHYSICS

THE BIG THEORY

MATTER VS ANTIMATTER

Matter vs Antimatter

From space drives to bananas – everything you need to know about the mysteries of antimatter

The very word 'antimatter' sounds like science fiction – on a par, perhaps, with antigravity. But antimatter is a perfectly real, well-established part of modern physics. It's true that it has some pretty dramatic possibilities, though. A gram of antimatter, if it came into contact with ordinary matter, would produce an explosion as big as a nuclear bomb. More usefully, because it's such an efficient source of energy, antimatter is the ideal candidate for future spacecraft propulsion. Surprisingly, however, there's nothing especially exotic about the nature of antimatter – it's a natural consequence of our current ideas about subatomic particles.

Most of the mass of an atom is contained in its nucleus, made of protons and neutrons. Orbiting around the nucleus are less massive particles called electrons. These are particularly important to us because of their role in electronics – each electron carries a negative electric charge (which is balanced by the positive charge on a proton). There are several other types of particle too, but they're usually only seen in high-energy physics experiments. All the particles that make up ordinary matter fall into two categories – 'baryons' like protons and neutrons, and 'leptons' like electrons.

This is starting to sound complicated, but it would get a lot more complicated if it wasn't for a fundamental principle of nature called conservation laws. These bring order to what otherwise might be complete chaos. When particles interact with each other – for example in the high-energy accelerators at CERN – certain

A magnetic 'trap' inside CERN's antimatter factory, capable of storing antihydrogen atoms

UNDERSTANDING PHYSICS

quantities are always conserved. Energy is one such quantity, and electric charge is another. It turns out that 'baryon number' and 'lepton number' are also conserved – and this is where we get onto the concept of antimatter.

A proton has a baryon number of +1 and a charge of +1. Theory also predicts its 'antiparticle' – with the same energy but a baryon number of -1 and charge of -1. It's called an antiproton. By the very same logic, the electron also has an antiparticle – called a positron – with positive electric charge and a lepton number of -1.

So what happens when, say, an antiproton meets a proton? You probably know the answer already, because it's the single best-known fact about antimatter. The positive and negative baryon numbers cancel out, as do the positive and negative electric charges and various other conserved quantities, until all that's left is the energy of the two particles. That's conserved too, but it's the only thing that's the same – not opposite – for both particles. So they disappear in a flash of energy – or more specifically, a gamma ray. That's an electromagnetic wave like light but with vastly greater energy, the same energy that just a moment ago was locked up inside the proton and antiproton. Called 'annihilation', it's the only process we know of that's capable of converting mass into energy with 100 per cent efficiency.

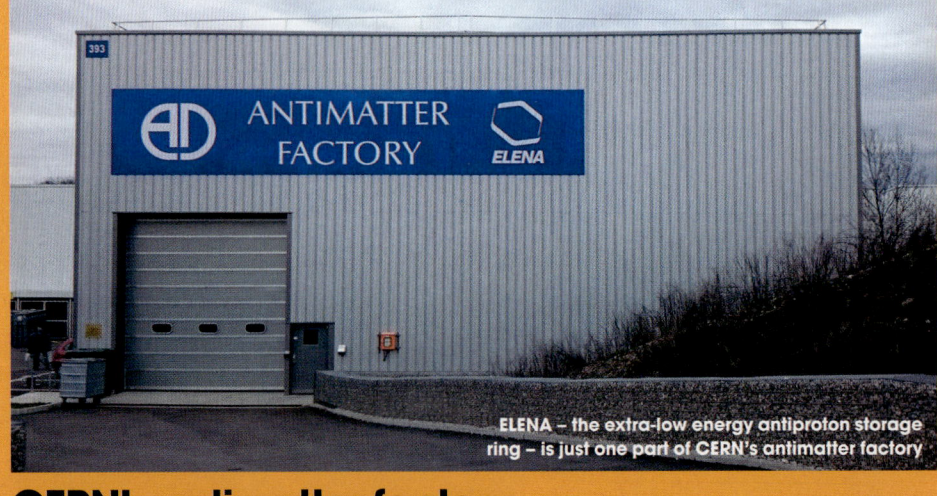

ELENA – the extra-low energy antiproton storage ring – is just one part of CERN's antimatter factory

CERN's antimatter factory

It's common for particle-antiparticle pairs to be created in accelerators at CERN and similar laboratories. The huge speeds mean collision energies are easily high enough. But the resulting antiparticles are short-lived, before they're annihilated in further collisions with ordinary particles. The purpose of CERN's 'antimatter factory' – the only facility of its type in the world – is to create antimatter that lasts long enough to study it properly. There are two tricks to this. The first is slowing the antiparticles to manageable speeds – the job not of an accelerator but a particle decelerator. Then the antimatter has to be confined somewhere so it can be studied, which is not an easy task, given that it would annihilate in contact with any kind of matter. The solution is to trap the antiparticles inside a strong magnetic field, where they can be used to make antihydrogen – an 'antiatom' consisting of a positron orbiting around an antiproton.

The reverse process is possible, too. Given enough energy, a particle-antiparticle pair can pop into existence from nowhere. With massive particles like protons and antiprotons, such pair production only takes place inside high-energy accelerators, or in exotic astrophysical processes. But the creation of electron-positron pairs is much more commonplace, occurring in certain types of natural radioactive decay here on Earth. And 'commonplace' is the right word: as we'll see later, even the humble banana is known to produce positrons. But antimatter produced in this way only survives for a fraction of a second. Almost as soon as it's created, it's going to encounter its normal matter counterpart and disappear in a tiny flash of gamma rays.

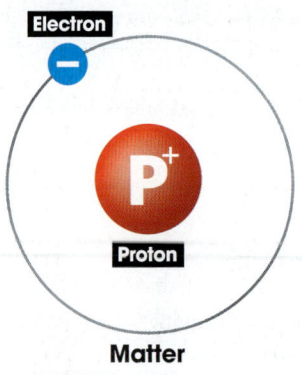

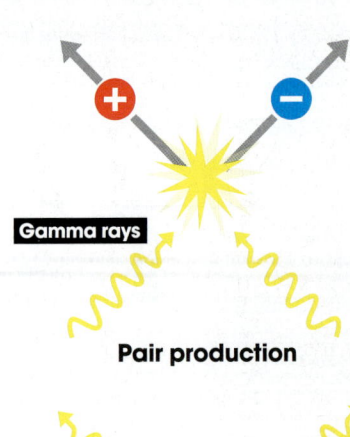

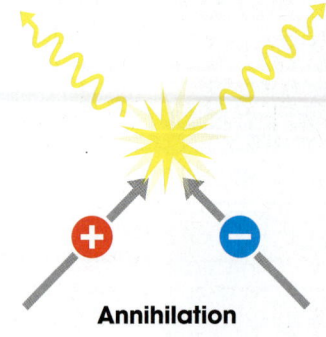

British physicist Paul Dirac predicted the existence of antimatter, based on theoretical equations, in 1928

Antihydrogen (right) looks like ordinary hydrogen (left), with an antiproton and positron replacing the proton and electron

An electron-positron pair can appear spontaneously with enough energy, or annihilate, producing the same energy

THE BIG THEORY

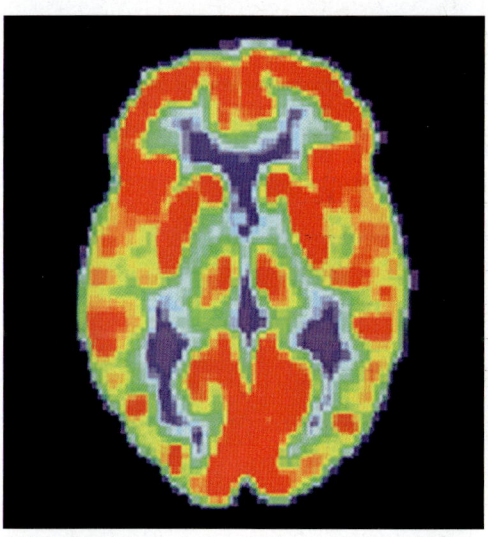

MATTER VS ANTIMATTER

Sources of antimatter

CERN may have the only antimatter factory, but there are other ways to make antiparticles

An interior view of a human brain, produced using a PET (positron emission tomography) scan

"Given enough energy, a particle-antiparticle pair can pop into existence from nowhere"

Everyday antimatter

Even atoms of ordinary matter sometimes produce antiparticles. The culprits are radioisotopes – atoms that are unstable because they have too few or too many neutrons. Several common substances contain a small proportion of isotopes like this, which decay to more stable forms by emitting high-energy particles. Usually these are ordinary matter – electrons in the case of beta decay, for example – but some radioisotopes also undergo 'beta-plus' decay, producing positrons instead. The positrons only last a fraction of a second before meeting electrons and annihilating to produce gamma rays. That's not as catastrophic as it sounds, because the energy of a single particle is tiny by everyday standards – which is fortunate, because there are positron-emitting isotopes inside your body. The commonest is potassium-40, which accounts for one in 10,000 of potassium atoms found in nature. It usually decays to regular beta-particles, but around 0.001 per cent of the time (1 in 100,000) the result is a positron.

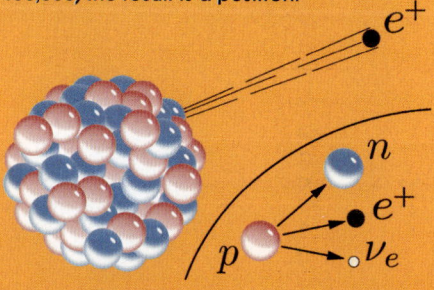

A schematic illustration of 'beta-plus' decay, in which an unstable nucleus emits a positron (e^+)

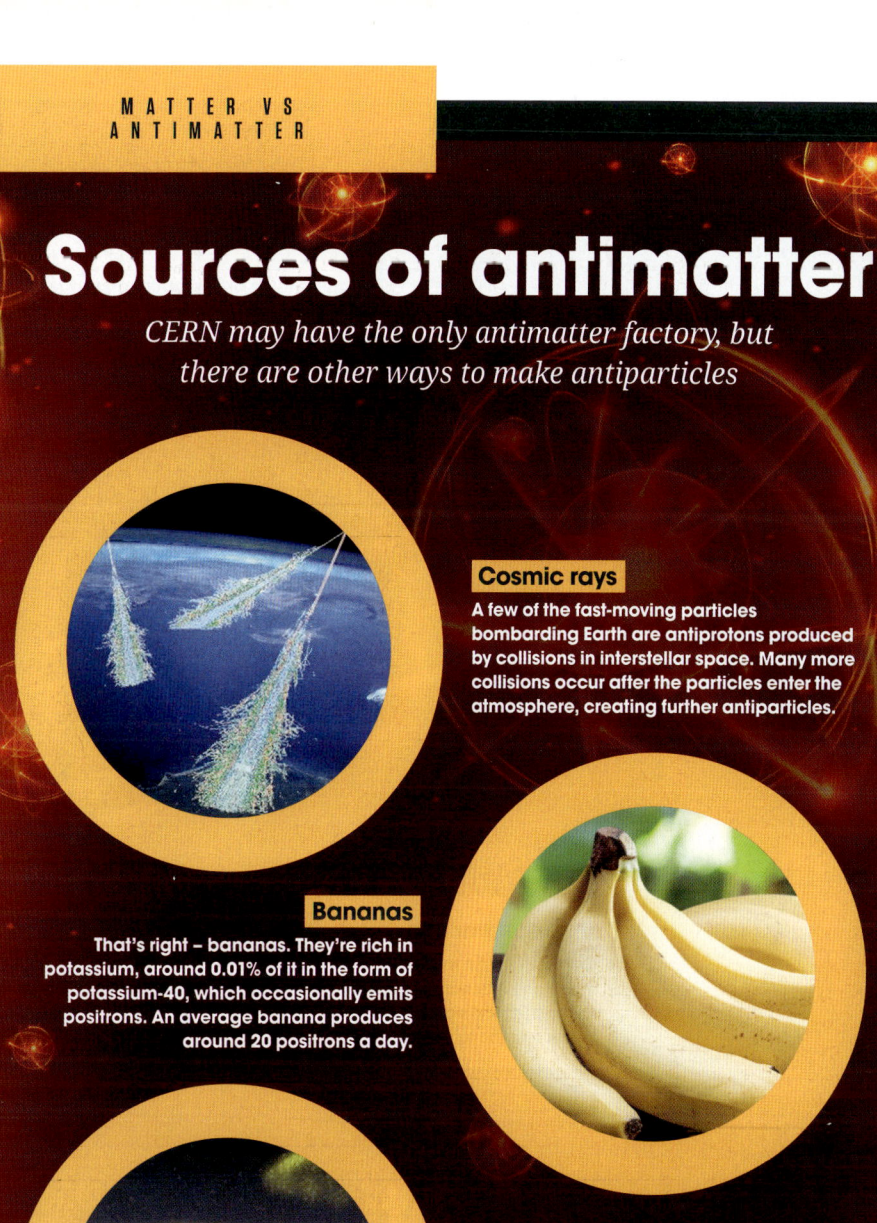

Cosmic rays
A few of the fast-moving particles bombarding Earth are antiprotons produced by collisions in interstellar space. Many more collisions occur after the particles enter the atmosphere, creating further antiparticles.

Bananas
That's right – bananas. They're rich in potassium, around 0.01% of it in the form of potassium-40, which occasionally emits positrons. An average banana produces around 20 positrons a day.

Thunderstorms
In 2011, a NASA satellite observed antimatter particles being created above thunderstorms on Earth – a result of the high-energy gamma rays that can be produced by lightning flashes.

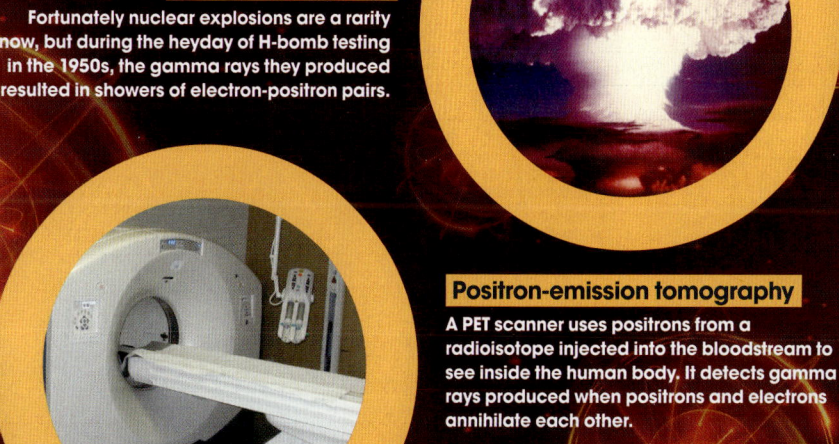

Nuclear explosions
Fortunately nuclear explosions are a rarity now, but during the heyday of H-bomb testing in the 1950s, the gamma rays they produced resulted in showers of electron-positron pairs.

Positron-emission tomography
A PET scanner uses positrons from a radioisotope injected into the bloodstream to see inside the human body. It detects gamma rays produced when positrons and electrons annihilate each other.

UNDERSTANDING PHYSICS

Payload
Depending on the application, the payload might be scientific instruments or even a crew module.

Deuterium tank
The basic propellant is deuterium – hydrogen with an added neutron – which is the ideal fuel for a nuclear fusion reactor.

Payload fraction
Antimatter is far more efficient than conventional propulsion, so the payload can be a larger fraction of the total mass.

A conceptual antimatter rocket

Antimatter space drives aren't complete science fiction – here's a design proposal that might really work

Positron source
This tank is filled with a radioisotope, such as krypton-79, which provides a continuous source of positrons.

Deuterium tank

Payload

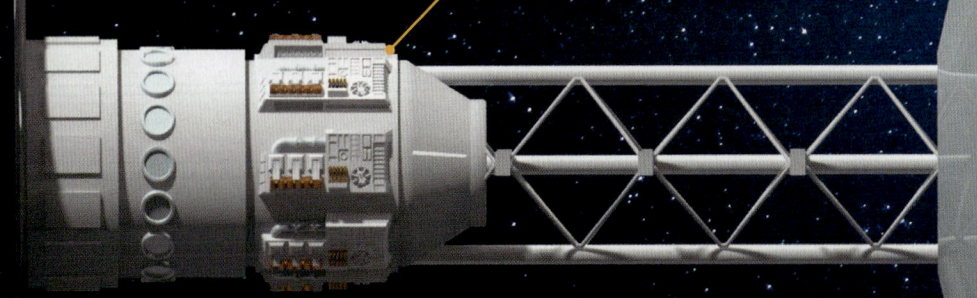

THE BIG THEORY

052

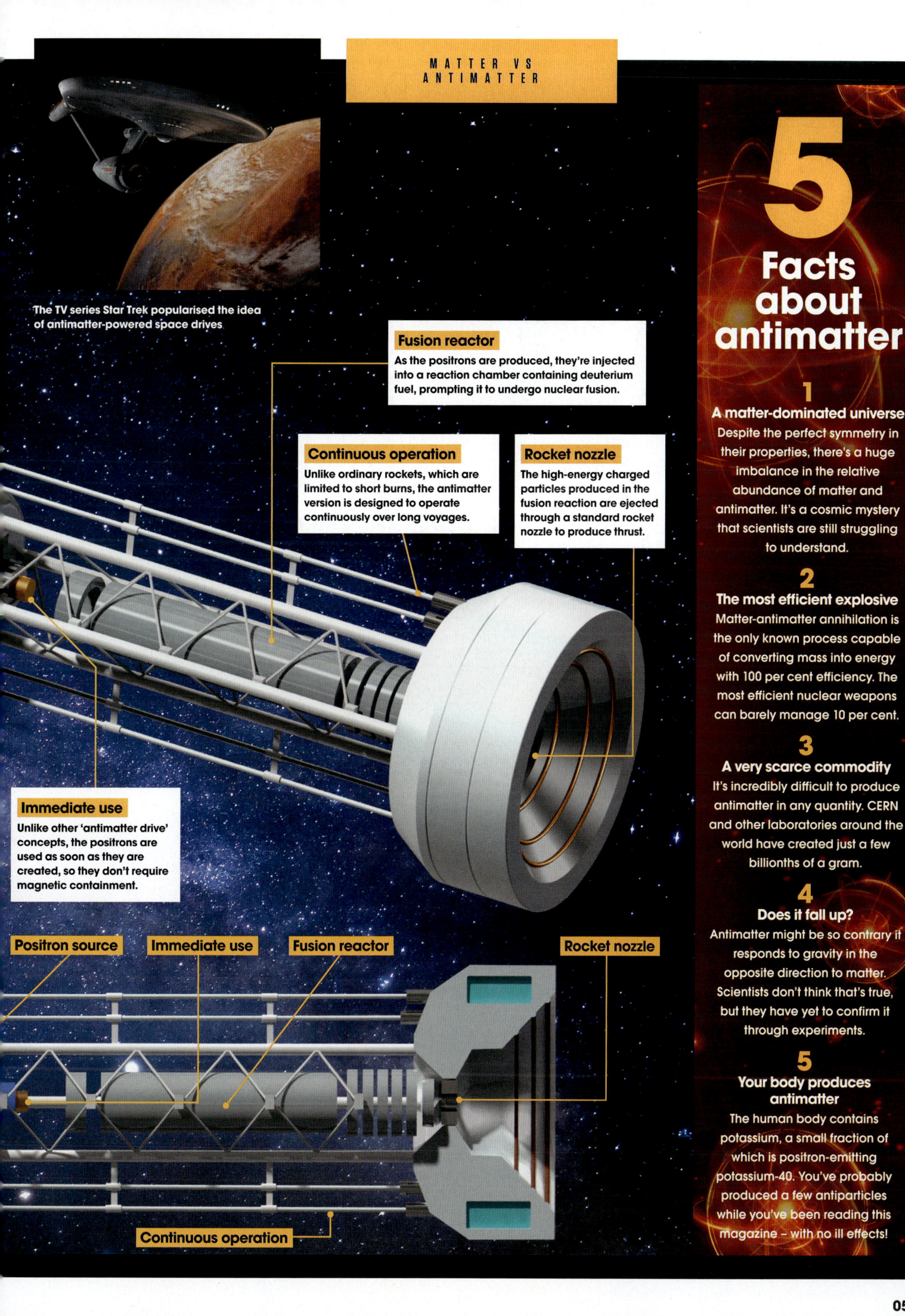

MATTER VS ANTIMATTER

The TV series Star Trek popularised the idea of antimatter-powered space drives.

Fusion reactor
As the positrons are produced, they're injected into a reaction chamber containing deuterium fuel, prompting it to undergo nuclear fusion.

Continuous operation
Unlike ordinary rockets, which are limited to short burns, the antimatter version is designed to operate continuously over long voyages.

Rocket nozzle
The high-energy charged particles produced in the fusion reaction are ejected through a standard rocket nozzle to produce thrust.

Immediate use
Unlike other 'antimatter drive' concepts, the positrons are used as soon as they are created, so they don't require magnetic containment.

Positron source | Immediate use | Fusion reactor | Rocket nozzle | Continuous operation

5 Facts about antimatter

1
A matter-dominated universe
Despite the perfect symmetry in their properties, there's a huge imbalance in the relative abundance of matter and antimatter. It's a cosmic mystery that scientists are still struggling to understand.

2
The most efficient explosive
Matter-antimatter annihilation is the only known process capable of converting mass into energy with 100 per cent efficiency. The most efficient nuclear weapons can barely manage 10 per cent.

3
A very scarce commodity
It's incredibly difficult to produce antimatter in any quantity. CERN and other laboratories around the world have created just a few billionths of a gram.

4
Does it fall up?
Antimatter might be so contrary it responds to gravity in the opposite direction to matter. Scientists don't think that's true, but they have yet to confirm it through experiments.

5
Your body produces antimatter
The human body contains potassium, a small fraction of which is positron-emitting potassium-40. You've probably produced a few antiparticles while you've been reading this magazine – with no ill effects!

UNDERSTANDING PHYSICS

Testing Hawking's theories

Which of Stephen Hawking's explanations turned out to be right?

Hawking presenting a lecture at George Washington University in 2008

Stephen Hawking was one of the greatest theoretical physicists of the modern age. Otherwise known for his appearances in popular media and his lifelong battle against a debilitating illness, his true impact comes from his brilliant five-decade career in science. Beginning with his doctoral thesis in 1966, his groundbreaking work continued nonstop right up until his final paper in 2018, completed just days before his death at the age of 76.

Hawking worked at the intellectual cutting edge of physics, and his theories often seemed bizarrely far out at the time he formulated them. Yet they're slowly being accepted into the scientific mainstream, with new supporting evidence coming in all the time. From his mind-blowing views of black holes to his explanation for the universe's humble beginnings, here are some of his theories that were vindicated, and others that are still up in the air.

Event horizon
The event horizon of a black hole is the point where nothing can escape its gravity. If a pair of particles is created near this boundary, one can get trapped inside.

What is Hawking radiation?
By applying quantum theory, Hawking realised the spontaneous emission of thermal radiation from black holes

Matter Creation
Most of the universe is made up of normal matter. But antimatter and matter are created in equal parts around active black holes.

This illustration shows the expansion of the universe, starting at the Big Bang around 13.8 billion years ago

The Big Bang wins
Hawking got off to a flying start with his doctoral thesis, written at a critical time when there was a heated debate between two rival cosmological theories: the Big Bang and steady-state. Both theories accepted that the universe is expanding, but in the first it expands from an ultra-compact, super-dense state at a finite time in the past, while the second assumes the universe has been expanding forever, with new matter always being created to maintain a constant density. In his thesis, Hawking showed that steady-state theory is mathematically self-contradictory. He argued instead that the universe began as an infinitely small, infinitely dense point called a singularity. Today, Hawking's description is almost universally accepted among scientists.

THE BIG THEORY

TESTING HAWKING'S THEORIES

Black holes are real

More than anything else, Hawking's name is associated with black holes – another kind of singularity, formed when a huge star undergoes complete collapse under its own gravity. These mathematical curiosities arose from Einstein's theory of general relativity, and they had been debated for decades when Hawking turned his attention to them in the early 1970s. His stroke of genius was to combine Einstein's equations with those of quantum mechanics, turning what had previously been a theoretical abstraction into something that looked like it might actually exist in the universe. The final proof that Hawking was correct came in 2019, when the Event Horizon Telescope obtained a direct image of the supermassive black hole lurking in the centre of giant galaxy Messier 87.

Black holes are so massive that even light can't escape their pull

Positive and negative
The particle that is trapped in the black hole has negative energy, while the particle that escapes has positive energy.

Hawking radiation
The particle of the pair that hasn't been trapped in the event horizon is emitted from the black hole as Hawking radiation.

This is a conceptual illustration of multiverse theory

5 theories scientists are unsure about

1 information paradox
Hawking believed that information about the basic properties of the material that made a black hole is stored in a cloud of zero-energy particles surrounding it. This is one of several hypotheses that have been put forward about black holes' lost material.

2 Primordial black holes and dark matter
Hawking was the first person to explore the theory behind black holes that were created soon after the Big Bang in depth. He said that these black holes might make up the mysterious dark matter that astronomers believe permeates the universe.

3 The multiverse
Hawking wasn't happy with the suggestion made by some scientists that any ludicrous situation you can imagine must be happening right now somewhere in one of an infinite number of universes. Instead he proposed a novel mathematical framework that rendered the universe finite.

4 Chronology protection conjecture
Hawking was bothered by the fact that Einstein's equations allowed backward time travel because he felt that it raised logical paradoxes that shouldn't be possible. He suggested that some currently unknown law of physics prevents these 'closed timelike curves' from occurring – his so-called 'chronology protection conjecture'.

5 Doomsday prophecies
In his later years, Hawking made a series of bleak prophecies concerning the future of humanity that he may or may not have been totally serious about. These range from the suggestion that the elusive Higgs boson might trigger a vacuum bubble that would gobble up the universe to hostile alien invasions and artificial intelligence (AI) takeovers.

"He worked at the intellectual cutting edge of physics"

AR zone SCAN HERE

55

UNDERSTANDING PHYSICS

High tech

58 Nuclear Power
Today's nuclear power stations versus a nuclear fusion future

64 Small science
How have microscopes revealed the tiny world around us?

70 How superconductors work so efficiently
Turn down the temperature on these materials to reveal their superpowers

72 Inside an atom smasher
This particle accelerator is solving the mysteries of the universe

HIGH TECH

70

72

UNDERSTANDING PHYSICS

Nuclear power

Investigate how today's nuclear power stations work and delve into the promise of nuclear fusion

HIGH TECH

The idea of harnessing energy from nuclear reactions to generate electricity is over 60 years old. Following a slow-down in the 1970s, nuclear power is now on the rise again, partially in response to concerns over the harmful effects of burning fossil fuels. Today's commercial nuclear reactors generate energy from the process of nuclear fission, and we'll investigate what that means, why it generates so much energy and how a nuclear power station works. However, while fission is a tried and proven technology, many scientists believe that the future is one of nuclear fusion. Over the next few pages, we'll take a look at that process to see how it differs from fission and how far we are from generating power from this potentially abundant energy source.

Chemical bonds contain a large amount of energy, which can be released by chemical reactions. Burning fossil fuels is a classic example, and the amount of energy that can be produced this way is evident if we think about how far a car can travel when a gallon of petrol is oxidised. But the amount of energy stored in chemical bonds is absolutely tiny compared to the amount of energy that is stored in the bonds between the protons and neutrons in the nucleus of an atom.

It is this energy that is released in the nuclear

NUCLEAR POWER

The ITER project aims to deliver the first large-scale fusion reactor by 2035

Fission vs fusion
Fission and fusion are opposite nuclear reactions, but both can generate energy

FISSION

Fission reaction
Fission is brought about when a neutron collides with a high atomic weight nucleus such as uranium-235.

Fission products
The result is two smaller nuclei – often barium and krypton in the case of uranium-235 fission – plus two or three neutrons.

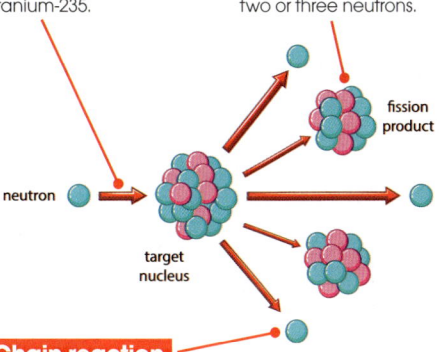

Chain reaction
More neutrons are generated than the one taken to initiate the fission. These reach other uranium-235 nuclei, resulting in a chain reaction.

FUSION

Deuterium and tritium
Deuterium nuclei contain one proton (yellow) and one neutron (purple). Tritium nuclei have an extra neutron.

Fusion products
Fusion generates a nucleus of a larger element (helium) and a neutron is emitted. Energy is released in the process.

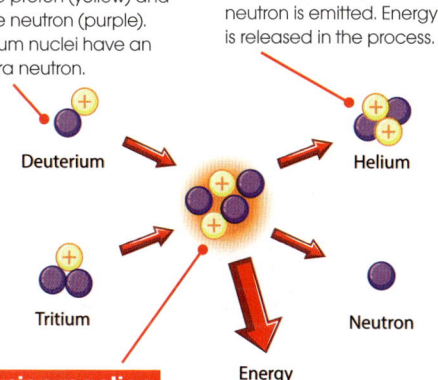

Fusion reaction
When brought together at over 100 million degrees Celsius, the deuterium and tritium nuclei undergo a nuclear fusion reaction.

---

reactions that take place in nuclear power plants, and the benefit compared to burning fossil fuels is staggering. Weight for weight, fission of nuclear fuel can produce 2-3 million times more energy than burning coal or oil.

Scientists first recognised the potential of nuclear energy in the 1930s and, while he was by no means the only researcher of note, Italian physicist Enrico Fermi has been acknowledged as the 'architect of the nuclear age'. In 1939, Fermi took up a position at Columbia University in the US, where he detected the release of energy from nuclear fission and, by 1942, had helped to build the world's first self-sustaining controlled nuclear chain reaction. However, political events were soon to alter the course of research into nuclear technology. With America involved in World War 2, Fermi was enlisted into the Manhattan Project where, together with other eminent scientists of the time, most notably Robert Oppenheimer, he would be instrumental in the development of the nuclear bomb.

Following the end of hostilities, attention returned to nuclear fission as an energy source for peaceful applications. The first commercial nuclear power station, Calder Hall in the UK, opened in 1956, generating 50 megawatts. Within just a few years, nuclear power plants were also operating in the US, Canada, France and the USSR. Today, there are about 450 operational stations in 30 countries.

Although the atom was once thought to be indivisible, nuclear fission involves just that, splitting an atom of a large atomic weight into two atoms of lower atomic weight. The element of choice, as used in nuclear power plants, is uranium – but not just any uranium. An element

> "Although the atom was once thought to be indivisible, nuclear fission involves splitting it"

Binding energy

The binding energy is the energy required to separate a pair of nucleons (ie protons and neutrons). It varies with the number of nucleons in the nucleus (the atomic weight), rising to a maximum between about 50 and 70 nucleons. Because both the large fissile (capable of fission) nuclei and the small fusile (capable of fusion) nuclei have a smaller binding energy than that of the nuclei they become during fission or fusion, energy is released in the reaction.

The shape of the graph also explains why more energy is available from fusion than fission. Due to the particularly steep curve for small numbers of nucleons, there is a large difference between the binding energy of fusile nuclei (2H and 3H) and that of the result of the fusion (4He).

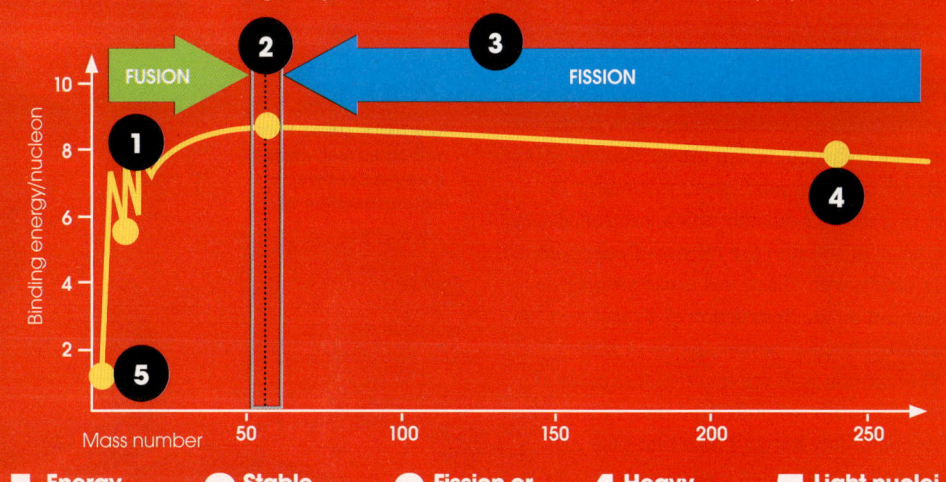

1 Energy release
Fusion causes light nuclei to become nuclei with a higher binding energy, releasing energy.

2 Stable region
Iron and the elements close to it in terms of atomic mass have tightly-bound nuclei.

3 Fission or fusion?
Lighter elements release energy by fusion, while the heavier elements release it by fission.

4 Heavy nuclei
Heavy nuclei are useful in fission reactions as they decay to nuclei with higher binding energies.

5 Light nuclei
Nuclei effective for generating energy by fusion have light nuclei, in other words not many nucleons.

UNDERSTANDING PHYSICS

is defined by the number of protons in its nucleus and for uranium this is 92, a number known as its atomic number.

However, elements can exist in several forms, known as isotopes, which differ in the number of neutrons in their nucleus. Uranium isotopes include uranium-235 (otherwise denoted as 235U) and uranium-238 (238U), where the number is the atomic weight, which is the sum of the number of protons and neutrons. Naturally occurring uranium is about 99.27 per cent uranium-238 and only 0.7 per cent uranium-235 – not particularly useful for energy generation because uranium-235 is the isotope that can undergo fission (uranium-238 cannot sustain a fission chain reaction). To be useful as a fuel, therefore, the concentration of the fissile uranium-235 has to be increased in a process called enrichment. Because the chemical properties of the two main isotopes of uranium are very similar, enrichment is a lengthy process in which the concentration of uranium-235 is increased in steps. The enriched uranium used for power generation has about three to five per cent uranium-235.

Fission of uranium-235 occurs when neutrons are fired at it. The neutron is initially captured by the uranium-235, but this makes it highly unstable, causing it to split into two other elements, releasing energy in the process. Fission of uranium-235 can give rise to a whole range of by-products, although isotopes of barium and krypton are two of the most common. Most of these by-products are highly radioactive in themselves, so they, in turn, also decay. Crucially, though, the fission reaction also releases two or three neutrons, which are then free to collide with other uranium-235 atoms, and so cause them to undergo nuclear fission. This gives rise to a chain reaction, which means that the fusion reaction, once initiated, is self-sustaining. In fact, in a nuclear reactor, unless controlled, this process will result in the release of energy much too quickly, with disastrous consequences, as evidenced by the destruction of a reactor at the Chernobyl nuclear power station in 1986.

The solution is to use a material capable of neutron capture without itself undergoing fission, most commonly boron. These materials are fashioned into so-called control rods and housed in the reactor core. By raising and lowering the control rods, the neutron flux can be controlled to allow the fission reaction to take place while preventing a runaway situation, an eventuality called criticality. They also allow an emergency shut down of the reactor.

Discussion of nuclear power stations inevitably leads to talk of the various types of reactor, with

1 Buffer fuel storage pool
Previously used fuel is stored here before reprocessing. The pool water is processed by a cooling system to recycle the water and prevent the spent fuel from overheating.

2 Reactor pressure vessel
The pressure vessel houses the core and the water that will transfer heat to the turbine.

3 Steam and water pipes
Pipework routes steam into the turbine and condensed water back to the reactor.

4 Containment vessel
The containment system prevents the release of radioactive substances into the atmosphere in the event of an accident.

Inside a nuclear fission plant

A tour of a power station based on GE Hitachi's Economic Simplified Boiling Water Reactor design

Refuelling machine
This robotic machine moves fuel rods into and out of the reactor during refuelling.

Fuel building
New fuel is stored here, as is spent fuel, which is stored underwater to reduce radiation risk.

Spent fuel is stored underwater to prevent radioactive discharge

Inclined fuel transfer system
The inclined fuel transfer system transfers new and used fuel between the fuel building and the containment vessel.

HIGH TECH

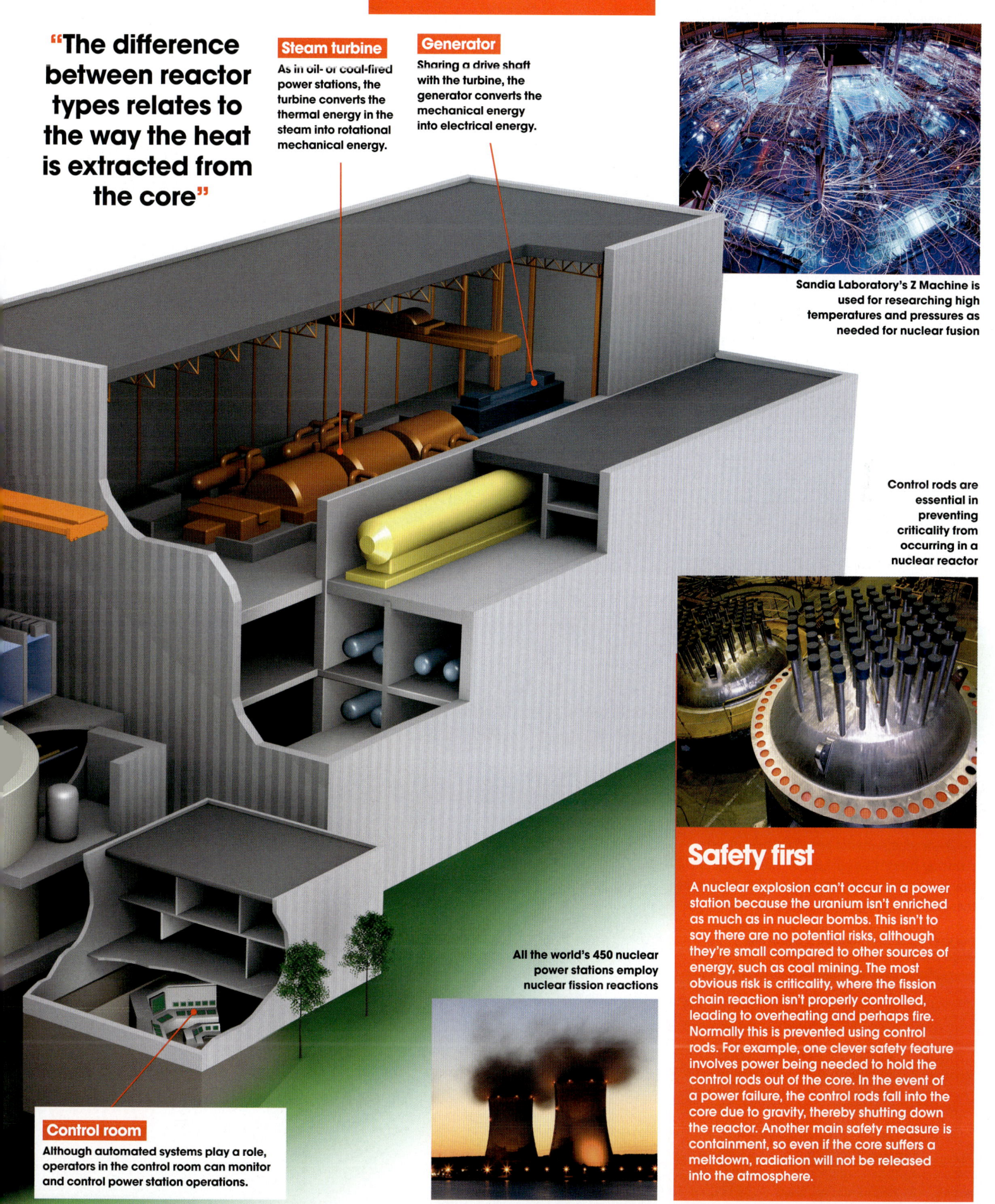

NUCLEAR POWER

> "The difference between reactor types relates to the way the heat is extracted from the core"

Steam turbine
As in oil- or coal-fired power stations, the turbine converts the thermal energy in the steam into rotational mechanical energy.

Generator
Sharing a drive shaft with the turbine, the generator converts the mechanical energy into electrical energy.

Sandia Laboratory's Z Machine is used for researching high temperatures and pressures as needed for nuclear fusion

Control rods are essential in preventing criticality from occurring in a nuclear reactor

Control room
Although automated systems play a role, operators in the control room can monitor and control power station operations.

All the world's 450 nuclear power stations employ nuclear fission reactions

Safety first

A nuclear explosion can't occur in a power station because the uranium isn't enriched as much as in nuclear bombs. This isn't to say there are no potential risks, although they're small compared to other sources of energy, such as coal mining. The most obvious risk is criticality, where the fission chain reaction isn't properly controlled, leading to overheating and perhaps fire. Normally this is prevented using control rods. For example, one clever safety feature involves power being needed to hold the control rods out of the core. In the event of a power failure, the control rods fall into the core due to gravity, thereby shutting down the reactor. Another main safety measure is containment, so even if the core suffers a meltdown, radiation will not be released into the atmosphere.

UNDERSTANDING PHYSICS

names such as the pressurised water reactor, the boiling water reactor and the Magnox or gas-cooled reactor bandied around. At the highest level, though, all nuclear power stations work in much the same way. The nuclear fission reaction generates heat, the heat turns water into steam and, from here on, things are the same as in a coal- or oil-fired power station. The steam drives a turbine, which in turn drives a generator that produces electricity.

The difference between the various reactor types relates to the way the heat is extracted from the core. In the boiling water reactor, the water that is heated to produce the steam is pumped through the nuclear reactor. In the pressurised water reactor, on the other hand, to prevent contaminated steam entering the turbine, there are two water circuits. The primary water flows through the reactor, which gives up its heat to the secondary water in a heat exchanger, the secondary water turning to steam and driving the turbine. The advanced gas-cooled reactor, as favoured in the UK, is similar except that the primary circuit, which transfers heat from the reactor to the water in the secondary circuit, uses carbon dioxide instead.

So much for the current state of play, but the Holy Grail of nuclear power is fusion rather than fission. As the name suggests, nuclear fusion is the opposite of nuclear fission – two atomic nuclei merging to produce an element with a larger atomic mass. Again this generates energy; the plentiful energy that the Earth receives from the Sun is the result of a massive nuclear fusion reaction. One of the Sun's fusion reactions, and the one that has been the subject of most research, occurs between two isotopes of hydrogen, namely deuterium (hydrogen-2) and tritium (hydrogen-3) to produce helium. Fusion produces much more energy than fission and the by-products are not as radioactive, thereby reducing concerns over nuclear waste, plus the raw materials are potentially plentiful. Yet despite these benefits, there are some serious challenges

The Rossing Mine in Namibia is one of the world's largest producers of uranium

to be met before fusion can form the basis for power generation. In particular, to initiate and maintain the reaction temperatures of over 100 million degrees Celsius are needed, and to hold the deuterium and tritium atoms together a magnetic field thousands of times that of Earth's own is required.

Burning fossil fuels generates greenhouse gasses, but nuclear fission – while producing about 11 per cent of the world's electricity without producing carbon dioxide – has its critics. While renewables will undoubtedly play an important role in the future, the potential benefits of fusion, should it ever come to fruition, can't be ignored. Currently, a project involving China, the European Union, India, Japan, South Korea, Russia and the United States is causing considerable interest. Called ITER, the aim is to produce the first operating large-scale fusion reactor by 2035, so watch this space.

The Wendelstein 7-X fusion reactor
The engineering wizardry behind the world's largest fusion reactor

Ports
No fewer than 253 ports provide access to the centre of the reactor for monitoring and regulating the reaction process.

Keeping cool
Heat insulating cladding, referred to as a cryostat, prevents the super-cooled magnets from heating up.

Cryo legs
The legs that support the structure have to bear a considerable weight of 725tn.

Non-planar coils
50 twisted coils form the super-conducting electromagnets. These produce the magnetic field to contain the plasma.

Five segments
Despite its bizarre and apparently random shape, the reactor is constructed from five almost identical segments.

Liquid helium
Liquid helium at -270 degrees Celsius allows the loops that form the magnets to be super-conductive.

HIGH TECH

NUCLEAR POWER

Central support structure
To operate as intended, the five segments that constitute the reactor must be accurately positioned by being fixed to a rigid central framework.

Comparing alternative fusion reactors

Most fusion reactor designs are toroids with external coils that generate a magnetic field needed to prevent the high temperature plasma from touching the reactor walls. But the magnetic field must have a twist.

Tokamaks
In tokamak reactors, a current flows through the plasma to create the twist.

Stellarators
In stellarator reactors, the whole machine is twisted to achieve a twist in the field.

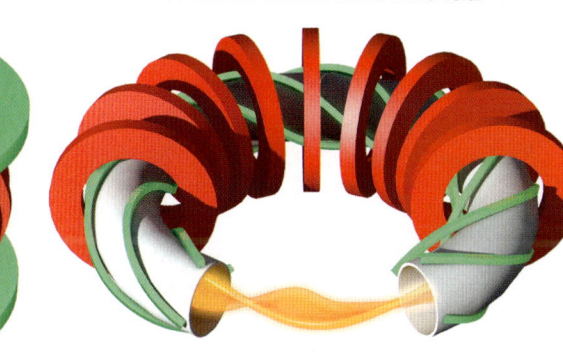

Vacuum vessel
To achieve a plasma the vacuum vessel maintains a pressure of less than a 100-millionth of atmospheric pressure.

Planar coils
These 20 coils allow for fine adjustments to be made to the magnetic field.

Plasma
The isotopes for fusion are heated to 100 million degrees Celsius, at which temperature they form a plasma.

The challenge of fusion

Research into nuclear fusion started decades ago and, for most of that time, commercial applications were thought to be 30 or 40 years away. So why are we getting no closer to a nuclear fusion power station? What's the challenge that's holding it at bay?

Unfortunately, there's no one single challenge but many. One of the most significant is that the necessary temperature is so high that large amounts of energy are needed. For many years experimental fusion reactors used more energy than they generated. A breakthrough came in 2014 at the Lawrence Livermore National Laboratory in the US, when a reactor generated 1.7 times more energy than it consumed. But the reactor was a small-scale device and the challenges are compounded as the technology is scaled up.

It's interesting to note that an aim of the ITER fusion reactor, scheduled for 2035, is to generate 500 megawatts but only use 50.

Lawrence Livermore's National Ignition Facility conducts fusion experiments using ultra-powerful lasers to heat and compress hydrogen fuel

> "The Holy Grail of nuclear power is fusion rather than fission"

UNDERSTANDING PHYSICS

Small Science

How have microscopes revealed the tiny world around us?

HIGH TECH

SMALL SCIENCE

What is the smallest thing you can see? A grain of sand? The lines of your fingerprints? Or perhaps, if you look really closely, the diameter of a human hair? Throughout most of human history, our eyesight was one of the biggest limitations on scientific research. Because we couldn't see cells or bacteria or atoms, we had no concept of these things, and it wasn't until the invention of the microscope in the 17th century that we started to understand the invisible world around us.

Scientists started to discover germs swarming in drinking water and miniature animals in lakes, and later they began to learn more about our own anatomy, finding taste buds and blood cells. Over the next century microscope technology boomed. Scientists worked to develop microscopes that were powerful enough to help diagnose cancer, seek out evidence at crime scenes, and, later in the 19th century, discover the building blocks of everything in our universe – atoms. From the humble beginnings of the simple microscope to the development of the first electron microscope, today we have far more advanced technology that can even view the space between atoms.

Microscopes are used to view and photograph very small objects that are invisible to the human eye. They can be categorised into two large groups: optical and electron. Optical microscopes are the ones you probably think of when you think of a microscope – they use a light source and a series of magnifying lenses so you can investigate your sample. This broad category is often used in diagnostic medicine and includes fluorescence microscopy, which observes fluorescence emitted by samples under special lighting, and laser microscopy, which uses laser beams to visualise samples.

Electron microscopes are even more complex, offering higher magnification and resolution. Instead of a beam of light, these pieces of equipment use a beam of electrons to create a projected image or record the bouncing back of electrons from the sample. There is also scanning probe microscopy, which includes atomic force microscopes that scan the surface of samples using a pyramid-shaped probe to map the surface of the specimen.

> "It wasn't until the invention of the microscope in the 17th century that we started to understand the invisible world around us"

WHY DO WE NEED ELECTRON MICROSCOPES?

When you are looking at something really small, if you have enough light your eyes can distinguish two points that are about 0.2 millimetres apart. This means the resolution of your eyes are about 0.2 millimetres. Light microscopes have much better resolution, and electron microscopes even more so. This is because electrons have much shorter wavelengths than white light, which has wavelengths of about 400 to 7,000 nanometres. The beams of electrons in an

Who invented the microscope?

Like many inventions, trying to work out who was the first person to build a microscope isn't very simple. Historians are undecided on if it was Hans Lippershey (the person who patented the first telescope) or the father-son spectacle makers Hans Martens and Zacharias Janssen. All three lived in the same town in Middelburg, Netherlands. These first microscopes were quite simple; they were just a tube with a lens placed at each end but could achieve up to 9x magnification.

While Zacharias Janssen and his father claimed that Lippershey stole the idea from them, a view that was backed up by letters from the Dutch diplomat Willem Boreel, Zacharias was known to be a very dishonest character who made a huge fortune from forging coins.

Hans Martens and his son Zacharias using an early microscope

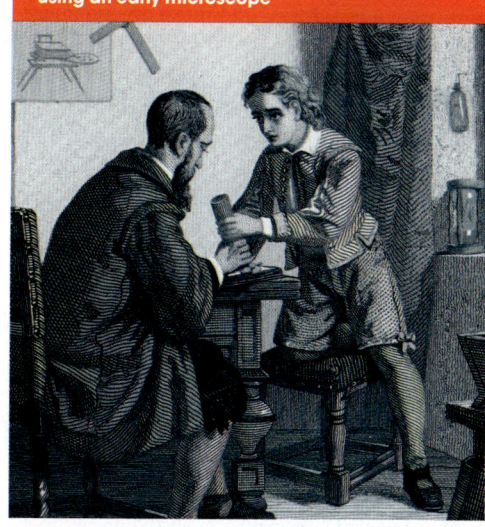

Under the microscope
Not even atoms can escape the glare of these tools

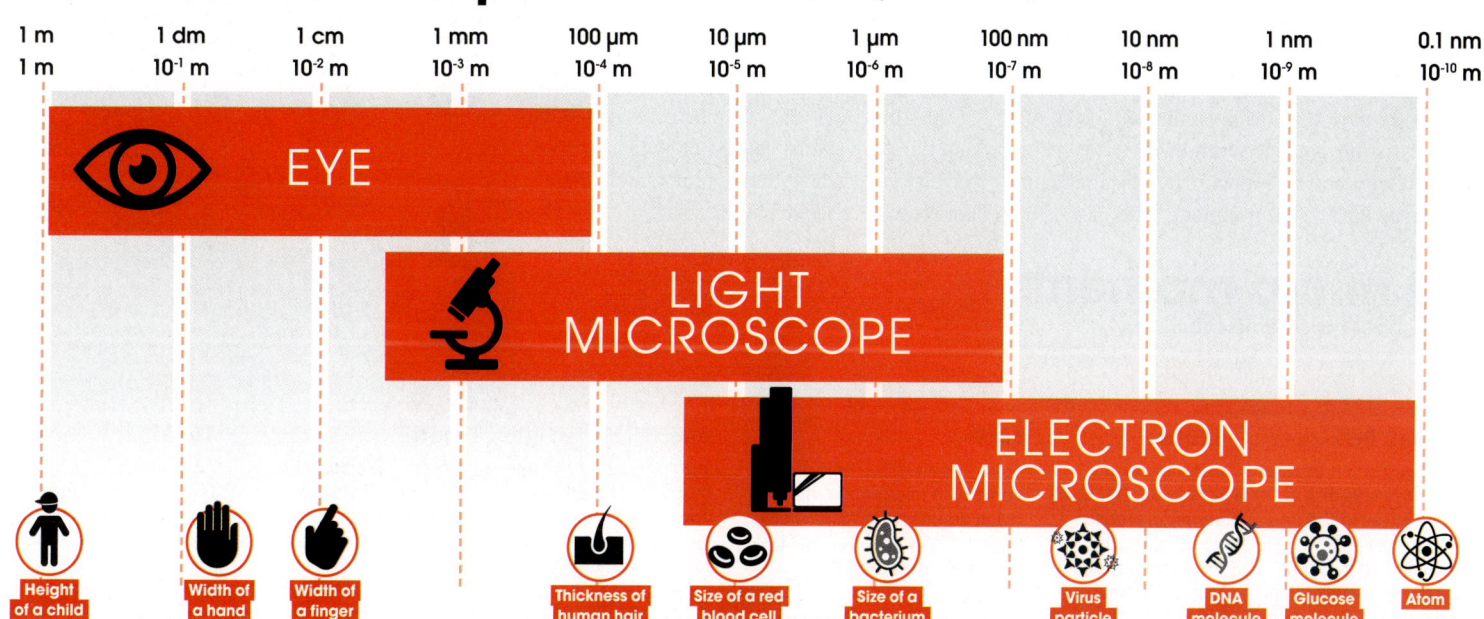

1 m	1 dm	1 cm	1 mm	100 µm	10 µm	1 µm	100 nm	10 nm	1 nm	0.1 nm
1 m	10^{-1} m	10^{-2} m	10^{-3} m	10^{-4} m	10^{-5} m	10^{-6} m	10^{-7} m	10^{-8} m	10^{-9} m	10^{-10} m

EYE

LIGHT MICROSCOPE

ELECTRON MICROSCOPE

- Height of a child
- Width of a hand
- Width of a finger
- Thickness of human hair
- Size of a red blood cell
- Size of a bacterium
- Virus particle
- DNA molecule
- Glucose molecule
- Atom

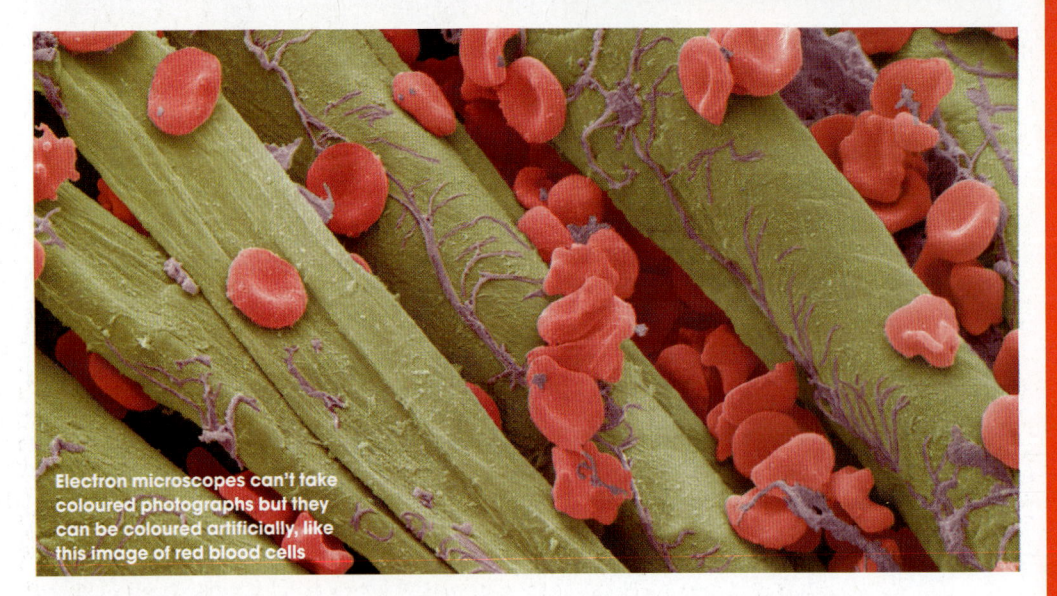

UNDERSTANDING PHYSICS

Electron microscopes can't take coloured photographs but they can be coloured artificially, like this image of red blood cells

electron microscope are nearer 0.1 nanometres. The smaller wavelength means less diffraction of light being scattered in random directions and as a result a less 'fuzzy' and more precise image is observed.

As scientists learn more and more about the microscopic world and our technology gets smaller, many structures of interest to research and development cannot be observed with light microscopy any longer. We require higher power and higher resolution to create things such as the tiny microchips inside our smartphones, and electron microscopes are becoming more popular.

MICROSCOPES IN DIAGNOSTICS AND CRIME SCENES

While technology relies on electron microscopes, many fields of biology are reliant on optical microscopes, particularly when it comes to identifying disease. Researchers use optical microscopes in diagnostics to observe samples. This is because diseases often leave a path of specific changes to the cells that can give a clue to what is happening to a patient, like the trademark dark dots inside malaria-infected cells, or the big gaps between brain tissue infected with bovine spongiform encephalopathy (known as BSE, or the infamous 'mad cow disease').

Optical microscopes are also utilised a lot in the field of forensics, where investigators diligently search for even the tiniest clues left at a crime scene and need to magnify evidence such as fingerprints or fibres from clothing.

THE FUTURE OF MICROSCOPES

There are many ideas and inventions that were created over the last decade that are still being developed for use in industry. At the forefront of pioneering work to improve microscope technology is the University of Manchester. Teams there have helped to develop a record-breaking optical microscope that has brought biologists a step closer to being able to view live viruses (which currently can only be viewed under an electron microscope).

Another project, which was launched in 2013 by the University of York, aims to combine the technology from both optical and electron microscopes into just one system in an attempt to overcome the challenges associated with both microscope types.

It might be hard to predict the technologies of the future that are yet to be constructed, but one thing we can be certain of is microscopes haven't yet reached their full potential. Who knows what else we will discover?

Five things scientists have discovered thanks to microscopes

1 Bacteria
Antonie van Leeuwenhoek discovered bacteria and protozoa swarming in water during the late 1670s. He sent beautifully detailed drawings of them to the Royal Society in London.

2 Cells
Plant cells were discovered by Robert Hooke in 1665. He was looking at dead cells from cork and named them 'cells' because he thought they resembled 'cellula' (the small rooms in monasteries).

3 Atomic nucleus
In the Geiger–Marsden experiments, scientists discovered that atoms contain a positively charged nucleus where most of its mass is concentrated by watching for the glow of alpha particles with a microscope.

4 Human genes
In 1995, Edward B Lewis, Christiane Nüsslein-Volhard and Eric Wieschaus find the genes involved in human development.

5 Sickle cell anaemia
In 1910 intern Ernest E Irons discovered the painful inherited condition sickle cell anaemia after he performed blood work on a student who had anaemia with odd crescent-shaped red blood cells.

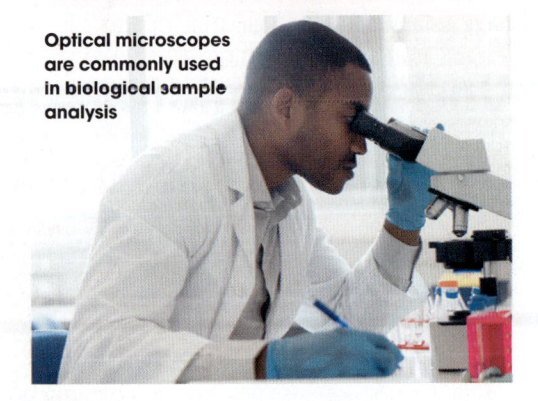

Optical microscopes are commonly used in biological sample analysis

Big micro moments
Microscopes have come a long way

750–710 BCE
The Nimrud lens is created from a rock crystal disc with a convex shape and used for burning (by concentrating the Sun's rays) or for magnification.

1200s
Using lenses in eyeglasses becomes common practice and single lens magnifying glasses become popular.

1619
Date of earliest description of a compound microscope after Dutch ambassador Willem Boreel sees one in London belonging to inventor Cornelis Drebbel.

1655
The first record of claims that Hans Martens and Zacharias Janssen invented the compound microscope in 1590.

1665
Robert Hooke publishes a collection of biological photographs in Micrographia and pioneers the word 'cell' for the shapes he finds in bark.

1673
Antonie van Leeuwenhoek improves the simple microscope in order to see biological samples. He later observes bacteria.

HIGH TECH

SMALL SCIENCE

Meet the microscopes

These machines use different techniques to let us see some of the smallest objects in our universe

Optical microscopes

Optical microscopes use light and a series of magnifying lenses to view specimens such as blood or tissue cells. They're probably the sort of microscope you used during science class at school. While they are the oldest microscope design, they remain vital in biological research and medical diagnostics.

Advantages
- Researchers can see the natural colour of the sample.
- Samples can be living or dead.
- Optical microscopes are not affected by magnetic fields.

Disadvantages
- The preparation to make a sample may distort specimen.
- Magnification is limited to 1500x.
- The resolving power (the distance needed to distinguish two points) for biological specimens is only around 1nm.

Scanning electron microscopes

Scanning electron microscopes use a beam of electrons that are scanned over the surface of a sample, which causes the production of secondary electrons, backscattered electrons and characteristic X-rays. These microscopes are held in vacuum chambers to prevent the electrons from hitting air molecules, and modern full-sized SEMs can provide a resolution between 1–20nm.

Advantages
- Minimal preparation of samples is required.
- Can provide detailed, three-dimensional and topographical imaging.
- Works fast and provides images within minutes.

Disadvantages
- Samples must be solid and able to tolerate vacuum pressure (not suitable for biological samples).
- Risk of radiation exposure due to the scatter of electrons from beneath the sample.
- Complicated and expensive, they are large and sensitive to electrical, magnetic and vibrational interference.

Transmission electron microscope

Transmission electron microscopes are the most powerful microscopes we have today. The electrons pass through the sample and are focused to form an image on a screen or onto a photographic plate. The faster the electrons hurtle down the microscope, the smaller the wavelength and the more detailed the image.

Advantages
- The most powerful microscopes, they can magnify by over 1 million times.
- Provide information on the element and compound structure of samples.
- Can determine shape and size as well as structure and surface features.

Disadvantages
- Samples must be 'electron transparent' (a thickness less than 100nm).
- Images are composed in black and white.
- Preparation of specimen is difficult and complex.

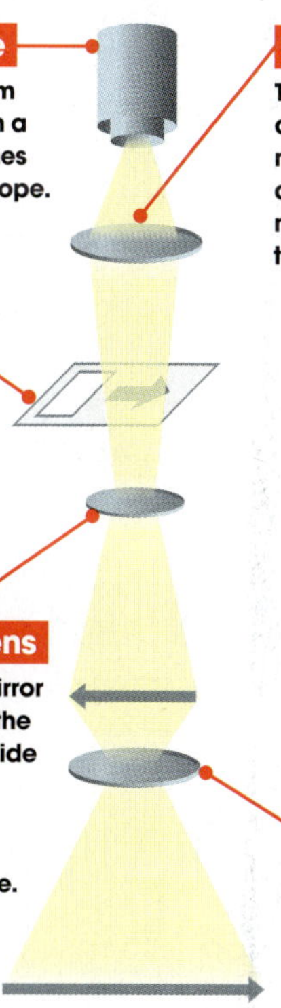

Light source — Natural light from the room or from a bright lamp shines into the microscope.

Specimen

Objective lens — Light from the mirror passes through the sample on the slide and hits the objective lens, illuminating a magnified image.

Angled mirror — The rays hit an adjustable angled mirror that changes the direction of the light, moving it upwards toward the sample.

Eyepiece lens — The eyepiece lens further magnifies the sample from the objective lens.

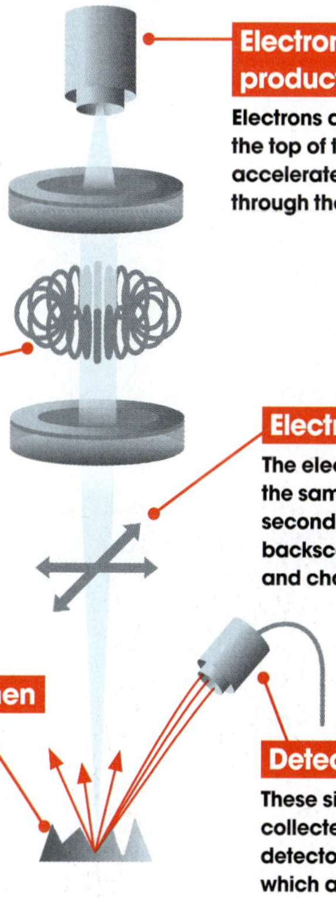

Electron production — Electrons are produced at the top of the column and accelerated down through the microscope.

Focused beam — Electrons pass through a combination of lenses and apertures to produce a focused beam.

Specimen

Electron interaction — The electrons interact with the sample, producing secondary electrons, backscattered electrons and characteristic X-rays.

Detection — These signals are collected by one or more detectors to form images, which are then displayed on the computer screen.

Cathode ray — A high-voltage electricity supply powers the cathode, which generates a beam of electrons.

Condenser lens — The first electromagnetic coil concentrates the electrons into a powerful beam that travels down the centre of the microscope.

Specimen

Projection lens — The projector lens magnifies the image, which becomes visible when the beam of electrons hits a fluorescent screen at the machine's base.

1873 — The Abbe sine condition is discovered by Ernst Abbe, a requirement that a lens needs to satisfy if it is to form a sharp image that is free of any distortions.

1951 — The field ion microscope is invented by Erwin Wilhelm Müller, making viewing atoms possible for the first time in history.

1953 — Professor of theoretical physics Frits Zernike receives the Physics Nobel Prize for inventing the phase-contrast microscope.

1967 — Erwin Wilhelm Müller builds on his original field ion microscope and creates the first atom probe, which allows the chemical identification of individual atoms for the first time.

1991 — The use of the Kelvin probe force microscope is published and is able to observe atoms and molecules.

2008 — Lawrence Berkeley National Laboratory installs a new $27-million microscope with a resolution of half of an angstrom. It remains the most powerful microscope.

UNDERSTANDING PHYSICS

Super STEM UK

A laboratory in Daresbury hosts some of Europe's most powerful microscopes

Some of the most powerful microscopes in the UK can be found in the countryside town of Daresbury, Cheshire. It's home to the UK National Facility for Advanced Electron Microscopy is funded by the Engineering and Physical Sciences Research Council (EPSRC). Here, researchers from all over the world come together to use the powerful microscopes that are kept at the facility. The newest model is the Nion UltraSTEM 100MC 'HERMES', also known as SuperSTEM 3, but the institute also houses the older models Nion UltraSTEM 100 (SuperSTEM 2) and the VG HB501 microscope equipped with a Mark II Nion Cs corrector (SuperSTEM 1).

These microscopes are a specialised type of TEM called scanning transmission electron microscopes (STEM), however, the 'HERMES' microscope can be used as a conventional transmission electron microscope (CTEM) as it is fitted with additional scanning coils to allow it to switch between different modes. The STEM machines produce images by using a focused beam of electron that scans across a thin sample in a raster pattern (horizontal, almost overlapping lines across a rectangular shape). The machines are so high resolution that they require an incredibly stable environment free from vibration, temperature fluctuations and electromagnetic and acoustic waves. This sensitivity can be demonstrated by clapping near SuperSTEM 2. The interference is immediately registered on the computer and jolts the atoms to one side.

While the SuperSTEM1 requires only a basic level of stability and atmospheric monitoring, the SuperSTEM 2 is shrouded in a heavy, thick curtain to reduce interference. The SuperSTEM3 is so sensitive that it must be operated from a separate room.

The SuperSTEM facility is keen to provide access for the global scientific community. Previous projects include investigating thermoelectric oxides for power generation and looking at molybdenum disulphide, a catalyst used in oil refineries, to remove harmful sulphur impurities in fossil fuels. Researchers from all fields are invited to apply to use the microscopes in small studies free of charge pending review by the scientists at the facility.

The Mesolens microscope

Modern optical microscopes have to compromise between the level of detail they can provide and the amount of sample they can show at a time. The giant Mesolens was created to overcome this limitation as it is designed to have both high resolution and a wide field of view. This powerful microscope lens is able to view both densely packed cells and the entirety of an embryo in one image, and it can magnify samples by up to 4x higher detail compared to conventional counterparts that produce the same magnification. The Mesolens has meant researchers are able to observe cells in situ complete with blood vessels and other surrounding tissue and can process a volume of sample 100x greater than when using a conventional microscope.

A culture of rat brain cells stained with fluorescent dyes including neurons (green), glial cells (red) and the nuclei (blue) of astrocytes

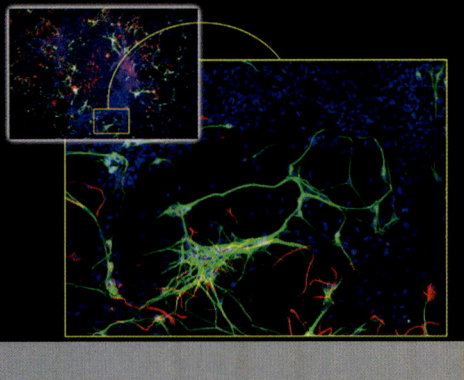

> "The Mesolens has meant researchers are able to observe cells in situ, complete with blood vessels and other surrounding tissue"

The SuperSTEM 1 is the oldest model of microscope at the facility and is an upgraded version of one of the machines built in 1970

HIGH TECH

SMALL SCIENCE

An interview with SuperSTEM's Demie Kepaptsoglou

How It Works interviews one of the scientists behind the project to uncover the universe's atomic secrets

There must have been some incredible things you've seen under these microscopes. What has been your favourite?
There's so many! Graphene, obviously. I remember the first time I looked at graphene. That was very cool because it's just a single atom thick and I was able to distinguish each atom. But also we have a collaboration with colleagues in Germany and they bring me meteorites that have travelled the universe – some of them are 4.5 billion years old. I was surprised to find out there is organic material in meteorites – there is this theory that it could be how the first organic matter came to Earth. There is a saying we have: 'We are investigating the universe, one atom at a time, but it might take us a while to get there.'

What is the importance of understanding the materials around us on an atomic level?
Do you remember the phone batteries that were exploding? These are batteries that are very, very small but are as powerful as a computer ten years ago. Obviously there was some fault in the production but it might not have been large at all because the products are so small now. We don't realise how much work and research goes into our everyday products.

Are there any advancements that you are excited to see in the future that will need electron microscopes?
I think drug delivery systems that will involve atoms and subatomic particles. There has been research into attaching magnetic nanoparticles to drugs so that they can use a magnet to guide the drug where they need it, [towards a] tumour or something.

Are nanoparticles dangerous to our health? Can you use electron microscopes to investigate this?
Yes, I was involved in an atmospheric study and they were collecting nanoparticles on the side of the road. They were determining what kind of nanoparticles were in our air and they found a lot of iron oxides coming from the brakes of cars. Understanding what things look like and how they act is very important to understanding the impact [of these very small particles] on health.

The SuperSTEM 2 is operated behind a curtain to protect experiments from interference due to temperature fluctuations or vibrations in the air

UNDERSTANDING PHYSICS

How superconductors work so efficiently

Superconductors may seem like perfectly ordinary materials, but turn down the temperature and their superpowers are revealed

Superconductors are metals – such as lead – or oxides which conduct electricity with no resistance. There's just one catch – to display their superpowers, they need to be kept at a frosty -265 degrees Celsius, or thereabouts.

Peer inside a chunk of lead and you'll see row upon row of neatly packed ions, bathed in a swarm of electrons. These loose electrons are what conduct electricity – set them into motion and you have an electrical current. At room temperature the lead ions vibrate away frenetically. From an electron's perspective, it's like trying to move across a crowded dance floor without spilling your drink. Constant collisions between electrons and ions convert electrical energy into heat – this is resistance.

Turn the temperature down a few hundred notches, however, and the ion vibrations subside, creating a stable lattice. Now as electrons flow through, a new effect comes into play: distortions in the lattice force them into pairs.

These unlikely unions trigger a weird quantum physics quirk: electron pairs throughout the material coalesce into a perfectly synchronised cloud, moving a bit like a school of fish. This means that the swarm of electrons can move through the lattice with no collisions, resulting in no resistance whatsoever.

Thanks to this astounding property, a huge current can be run through a superconductor without it overheating. This means that they can create incomparably powerful electromagnets. These are currently used in MRI scanners, supercomputers, particle accelerators like the Large Hadron Collider and levitating maglev trains.

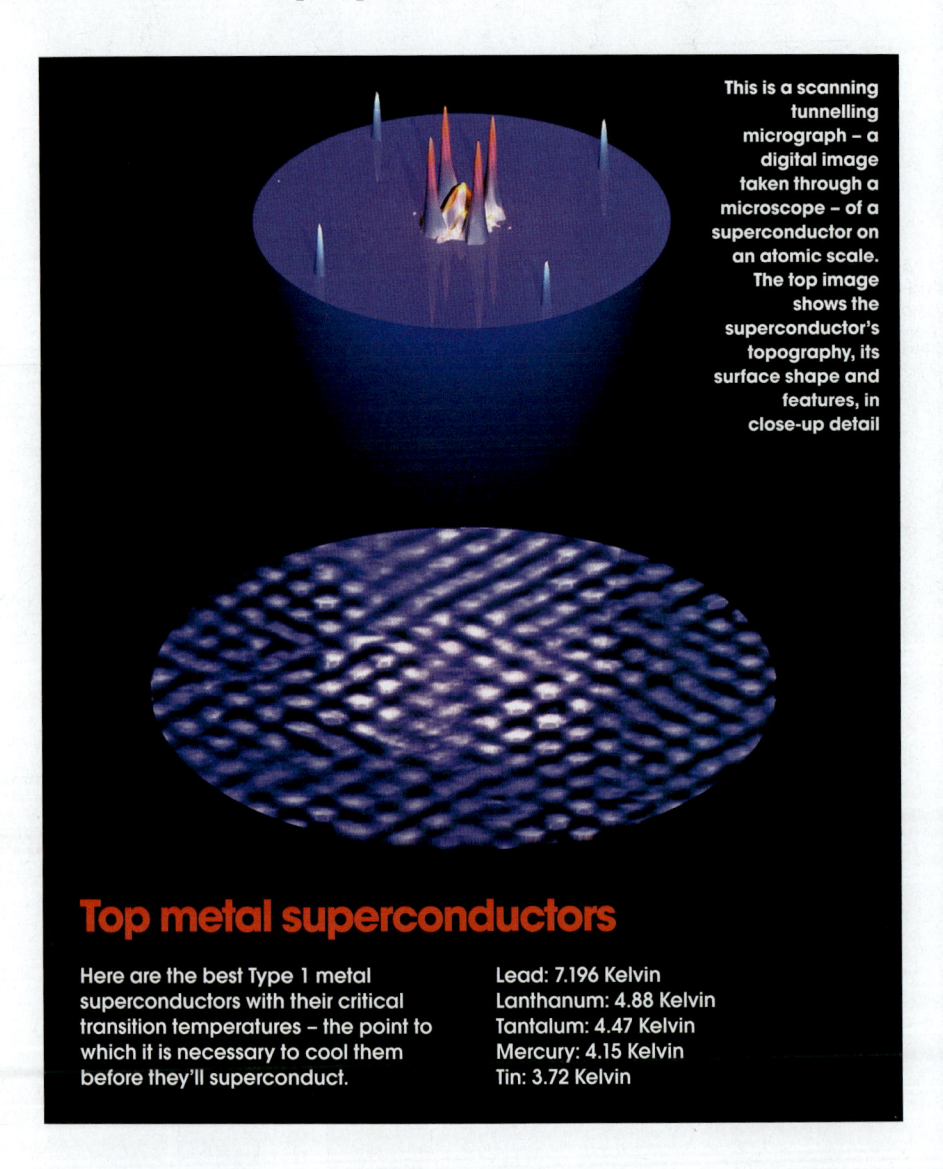

This is a scanning tunnelling micrograph – a digital image taken through a microscope – of a superconductor on an atomic scale. The top image shows the superconductor's topography, its surface shape and features, in close-up detail

Top metal superconductors

Here are the best Type 1 metal superconductors with their critical transition temperatures – the point to which it is necessary to cool them before they'll superconduct.

Lead: 7.196 Kelvin
Lanthanum: 4.88 Kelvin
Tantalum: 4.47 Kelvin
Mercury: 4.15 Kelvin
Tin: 3.72 Kelvin

Superconductor evolution

How It Works takes a journey through the last century to see just how far superconductors have come

1911
Absolute zero
Dutch physicist Heike Kamerlingh Onnes and and his team create temperatures of just above absolute zero and discover that mercury is a good superconductor.

1933
Levitation
Meissner and Ochsenfeld discover the Meissner effect: the uncanny ability of superconductors to repel magnetic fields and cause magnets to levitate.

1935
Brothers London
Fritz and Heinz London reconcile superconductor theory to show that zero resistance and the Meissner effect stem from the same phenomenon.

HIGH TECH

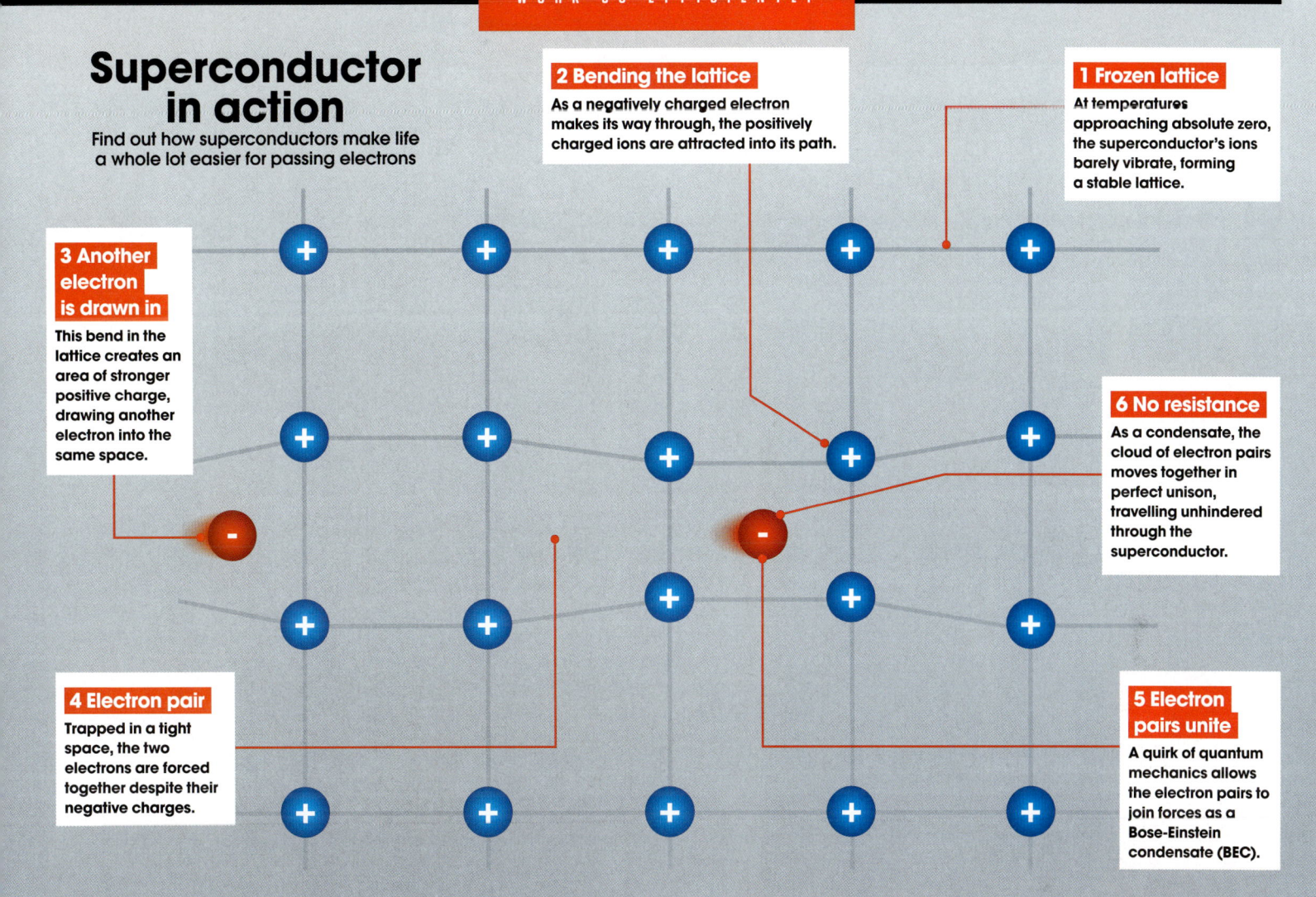

HOW SUPERCONDUCTORS WORK SO EFFICIENTLY

Superconductor in action
Find out how superconductors make life a whole lot easier for passing electrons

1 Frozen lattice
At temperatures approaching absolute zero, the superconductor's ions barely vibrate, forming a stable lattice.

2 Bending the lattice
As a negatively charged electron makes its way through, the positively charged ions are attracted into its path.

3 Another electron is drawn in
This bend in the lattice creates an area of stronger positive charge, drawing another electron into the same space.

4 Electron pair
Trapped in a tight space, the two electrons are forced together despite their negative charges.

5 Electron pairs unite
A quirk of quantum mechanics allows the electron pairs to join forces as a Bose-Einstein condensate (BEC).

6 No resistance
As a condensate, the cloud of electron pairs moves together in perfect unison, travelling unhindered through the superconductor.

A chilled superconductor repels magnetic fields, allowing it to levitate a magnet

The potential of superconductivity

Despite their impressive abilities, most current superconductor technologies remain chained to high-tech science laboratories, burdened by bulky, energy-greedy and very expensive cooling systems in order to function.
Scientists have set their sights on creating a superconductor that works at room temperature and pressure, which could bring cutting-edge technologies into all of our day-to-day lives. Inexpensive, portable MRI scanners could drastically improve healthcare, while superfast maglev trains would zip up and down the country, reducing travel times.
Replacing our inefficient electrical grids with superconducting cables would slash our electricity bills too. It could also give renewable energies – such as wind farms, which are often located great distances from our cities – a much-deserved boost. Elsewhere, superconductor-enabled electronics could see smaller, faster computers hit the high street.
While physicists have managed to create superconducting materials operational at a temperature of 15 degrees Celsius, they require extremely high pressures, approaching those found at the Earth's centre. Many still believe that the Holy Grail of truly room-temperature superconductors is achievable – it's just a matter of time and patience before we discover it and utilise it in new technology.

1957 BCS
Bardeen, Cooper and Schrieffer propose the BCS theory of superconductivity, explaining electron pairing. It earns them a Nobel Prize.

1986 Hot stuff
Bednorz and Müller discover the first 'high-temperature' superconductor, which works its magic up to -243.15 degrees Celsius.

2020 Hotter stuff
A metallic compound made of hydrogen, carbon and sulphur exhibits superconductivity at 15 degrees Celsius – but at extreme pressures.

UNDERSTANDING PHYSICS

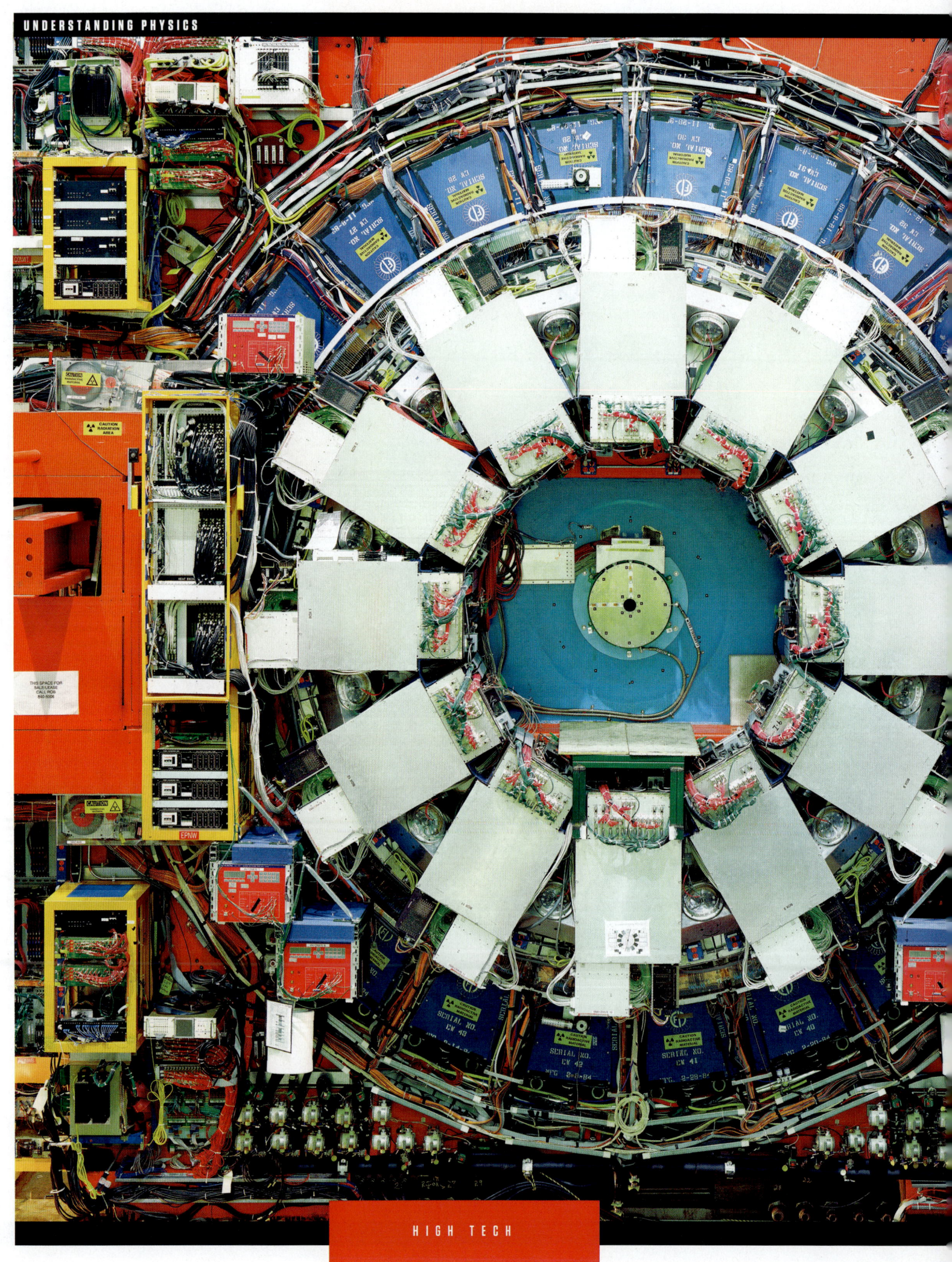

HIGH TECH

INSIDE AN ATOM SMASHER

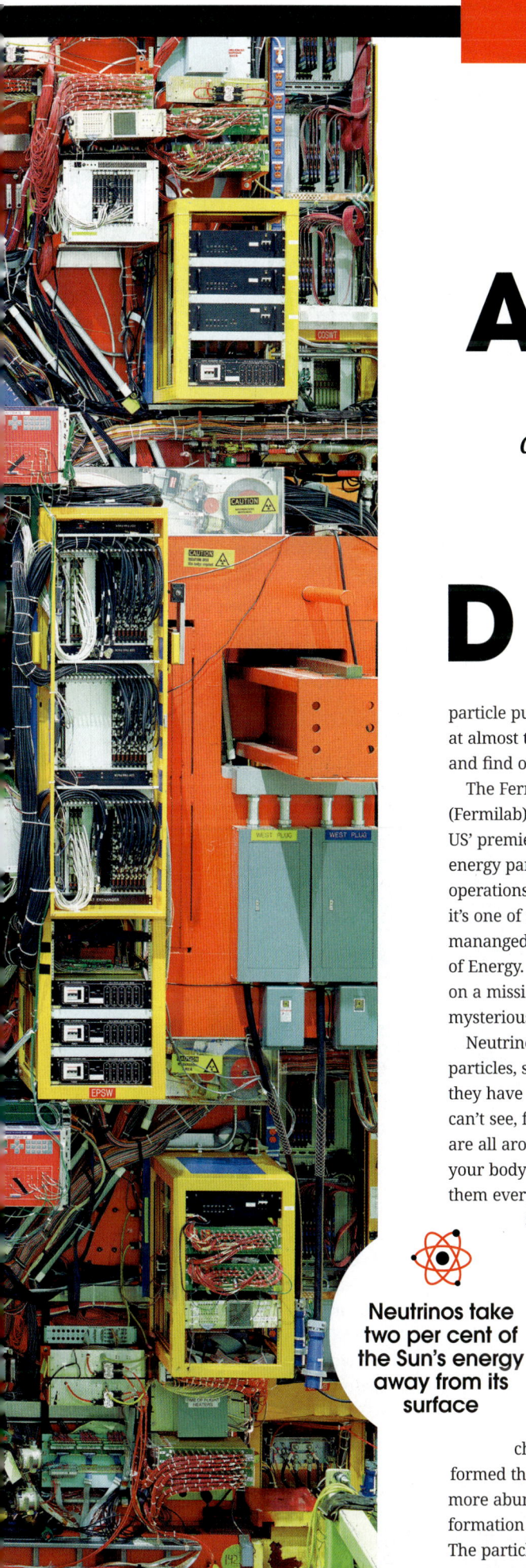

Inside an Atom Smasher

Take a trip into a particle accelerator and discover the experiments that are solving the mysteries of the universe

Deep underground in the US Midwest, groundbreaking projects are utilising advanced particle technology to examine tiny subatomic matter. From a subterranean facility, powerful particle pulses are sent across states every second at almost the speed of light. The reason? To try and find out why we exist. Welcome to Fermilab.

The Fermi National Accelerator Laboratory (Fermilab) near Chicago, Illinois, is the US' premier laboratory for high-energy particle physics. It began operations on 15 June 1967 and it's one of 17 national laboratories mananged by the US Department of Energy. Its staff are currently on a mission to locate and study mysterious particles called neutrinos.

Neutrinos are subatomic elementary particles, similar to an electron or proton, except they have less mass and no electrical charge. You can't see, feel, hear or smell them, but neutrinos are all around us. And they're passing through your body at a rapid rate – 100,000 billion of them every second actually, give or take. Solving their mysteries could potentially increase our knowledge of the origins of matter. Neutrinos can't be seen with the naked eye, but they could be vital to how the universe works. It's thought there were equal amounts of matter and antimatter (a partner particle with the same mass but opposite electric charge) shortly after the Big Bang that formed the universe. Then matter became much more abundant than antimatter, allowing for the formation of atoms, stars, planets and humans. The particle accelerators at Fermilab can send out both neutrinos and antineutrinos, their antimatter counterpart, so if a difference is found in how neutrinos and antineutrinos behave, it may help to explain how the universe evolved to end up without antimatter.

Neutrinos are created naturally in huge nuclear reactions in the Sun, or when a star supernovas, but can also be made in nuclear power plants and using particle accelerators, like at Fermilab. To try and analyse these rare particles, various projects have been set up at Fermilab. The first was DONUT in the late 1990s, which was followed by MINOS in 2005. The NOvA Neutrino Experiment began in 2014 and upped the ante – it's one of the largest experiments of all time.

The particle accelerator used in NOvA fires a beam of protons to a detector more than 800 kilometres away in Ash River, Minnesota. The particles don't need a tunnel to travel through as they simply go straight through the Earth. This 14,000-ton detector is filled with light-conducting fibres that record

40,000
Cosmic rays go through the detector every second

Neutrinos take two per cent of the Sun's energy away from its surface

The Fermilab laboratory is located on a 27.5 square-kilometre site

UNDERSTANDING PHYSICS

100 thousand billion
Neutrinos shot from the Fermilab every second

Fermilab is also home to many species of birds and insects, as well as coyotes and bison

The main injector for the particle accelerator used in NOvA

Fermilab isn't the only place to have a neutrino detector – there's one at CERN and even one at the South Pole

The NOvA far detector is the biggest free-standing plastic object in the world

the energy from neutrinos colliding with other particles. Along with the fibres, there are 344,000 cells of reflective plastic that are packed with 11 million litres of clear liquid that illuminate when particles come into contact. The facility uses cryogenic technology to keep the machines at -15 degrees Celsius, its optimum operating temperature. The detector is so gargantuan that a unique transport machine was required to move the 28 200-ton blocks that make it.

NOvA analyses how neutrinos change or 'oscillate' into different types. Neutrinos leave almost no trace and rarely interact with each other or other particles. The particle accelerator shoots protons, which then slam at very high energies into the target at Ash River. This creates shortlived particles that then decay to

> "NOvA analyses how neutrinos change or 'oscillate' into different types"

produce neutrinos. When neutrinos collide with other particles, the traces of the interactions are received by a detector, are examined by physicists and compared to previous statistics. Scientists are looking for trends in the data to decipher what neutrinos do and how they act.

One of the key breakthroughs that physicists have made is that there are different types, or 'flavours', of neutrino – each named after which electric-charged particle it collides with. Neutrinos

are from the lepton family of particles, and like leptons there are three types – muon neutrinos, electron neutrinos and tau neutrinos. Electron neutrinos are produced when a neutrino slams into an electron, for example. As neutrinos blast through the beam, they change between the three types frequently. Starting off as a muon neutrino, they often oscillate to electron and tau neutrinos. Neutrino oscillation is like a piece of fruit changing into a vegetable when you leave the supermarket, or this magazine changing into a book before you get home. Understanding why this happens will be key to understanding the nature of neutrinos.

Is there an endgame for NOvA? It has helped to increase scientific knowledge of neutrino oscillation and furthered the search

HIGH TECH

074

Q&A

Working on NOvA with Fermilab senior scientist, Peter Shanahan

Peter is one of Fermilab's 1,750 employees. The facility works with more than 50 countries on many collaborative experiments.

How did you become interested in studying neutrinos?
Towards the end of my postdoctoral research on a different topic, the field of neutrino research had just had a breakthrough, with the discovery of neutrino flavour oscillations. I was excited by the potential for investigation in this newly energised field. I ended up getting a position at Fermilab to work on the very first experiment to do that in the US, called MINOS, and I have been working on neutrinos ever since.

What's working with this tech like?
Most of the hands-on work is done by technicians, graduate students and postdoctoral researchers. This includes swapping out the occasional broken sensor or electronics card and using computers to analyse the data we take.

As you get more senior, your work tends to get a bit more hands-off. We keep an eye on the systems that monitor the detector, and data-taking, including performing checklists to make sure everything is running and that we are taking good data.

What is most exciting about your job?
Our experiments often take decades to plan, build, operate and produce final results. Along the way there are many exciting milestones. For example, we do extensive simulations of a planned experiment. In that process, we see for the first time how well a new detector technology should be able to perform in tracking particles, based on detailed simulations of the detector and the particles going through it. The first time a new detector sees an unmistakable cosmic ray particle is always exciting.

Then, your experiment produces new physics results – perhaps by providing a new measurement that is more precise than preceding ones, or by answering a question that no one had been able to answer before.

INSIDE AN ATOM SMASHER

1.6km of solid rock will protect DUNE from cosmic rays

A technician examines a section of Fermlab's Collider Detector

for a fourth neutrino type. University College London and Sussex University in the UK are both collaborating with the project by helping to analyse the oscillations. NOvA will collect data until 2024, ten years after it was first switched on, when it will be replaced by an all-new project called DUNE.

Work on DUNE, or 'Deep Underground Neutrino Experiment', began in July 2017. The project will be the largest international science experiment to take place in the US. It will be the strongest particle beam in the world, sending particles 1,300 kilometres to the Sanford Underground Research Facility in Lead, South Dakota. Excavation for the Long-Baseline Neutrino Facility, which will house it, begins later this year and is planned to be up and running by 2022. The European Organization for Nuclear Research, or CERN, which houses the Large Hadron Collider, has its own, slightly smaller detector, which went online in September 2018. A second detector is also on the way.

DUNE will benefit from a significant upgrade to Fermilab's accelerators, with the Proton Improvement Plan II (PIP-II). PIP-II will provide a new particle accelerator that will generate a proton beam with 60 per cent more power than before. The mechanism will be made from superconducting materials with no electrical resistance, resulting in even more power for a lower cost, and there will be more neutrinos to study than ever. DUNE will also have more sensitive detectors, using liquefied argon operating at -185 degrees Celsius. By 2026 the project will be fully operational.

Physicists at Fermilab will continue to study neutrinos to try and unlock the secrets they hold. Neutrinos can travel vast distances over the universe quickly, as few other particles, including those in magnetic fields, interfere with them. Because they are so difficult to locate, neutrinos could expose aspects of nature that are unknown to science, and potentially reveal the reasons why the universe is made of matter. We're only just beginning to understand the potential wonders of neutrinos and, as technology improves and our knowledge grows, there could be some startling revelations around the corner.

UNDERSTANDING PHYSICS

How to make a neutrino beam
Under the hood of the highest-intensity neutrino beam in the world

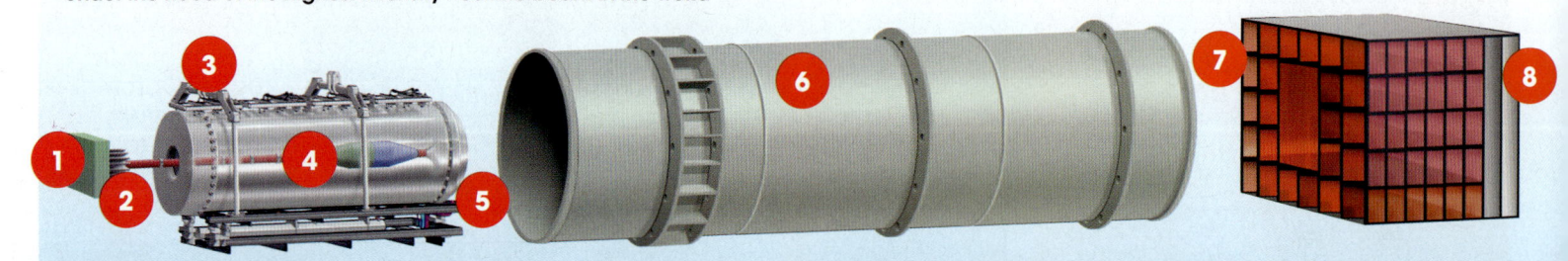

1 Feeding the mechanism
A proton beam travelling at nearly the speed of light is directed into the mechanism.

2 Break it down
In the graphite target, pions and kaons are produced from the energy of the protons slamming into neutrons and other protons.

3 You spin me right round
The power is generated using a circular accelerator system, a set of rings 3.3km in circumference.

4 Creating the beam
A magnetic focusing horn uses a magnetic field to concentrate the particles into a beam, eliminating interference from other matter.

5 It's getting hot in here
The powerful beam increases the temperature of the horn by 370 degrees Celsius, so a water and wind system keeps it cool.

6 Decaying away
In the decay pipe, the pions and kaons decay into even smaller particles called muons, as well as neutrinos.

7 Completing the process
The muons and neutrinos strike a beam absorber, which blocks the former but cannot stop the latter.

8 The neutrino beam
Now the beam is solely neutrinos, giving the best chance for their rare interferences with matter to be picked up by the detector.

DUNE uncovered
Inside the subterranean journey of neutrinos from Fermilab to Sanford

Interference from space
Neutrinos can also enter the detector naturally from the atmosphere and even from star supernovas.

Straight through the Earth
No tunnel is required, such is the power of the moving particles, and the beam simply passes through solid rock.

15 billion years old
The estimated age of many of the neutrinos found in the NOvA experiment

SANFORD UNDERGROUND RESEARCH FACILITY

PARTICLE DETECTOR

1,300 KM

EXISTING LABS

UNDERGROUND PARTICLE DETECTOR

The beam widens
As the beam heads towards South Dakota, it widens in the same way as a ray of light.

Journey's end
Scientists analyse 3D images to locate neutrino tracks and potentially any previously unseen data.

Incoming neutrinos
A detector produces readings of the particles including traces of neutrinos.

V_e
V_μ
V_τ

1,600 1,400 1,200 1,000 800 600 400 200 0km
Probability of detecting electron, muon and tau neutrinos

Incoming beam: 100% muon neutrinos

HIGH TECH

INSIDE AN ATOM SMASHER

Tiny collisions
What happens when a neutrino smashes into another particle in NOvA?

Beam entry
The beam sent from the particle accelerator speeds through the Earth.

Impact
The interaction of particles including a neutrino causes a collision that is registered by the detector.

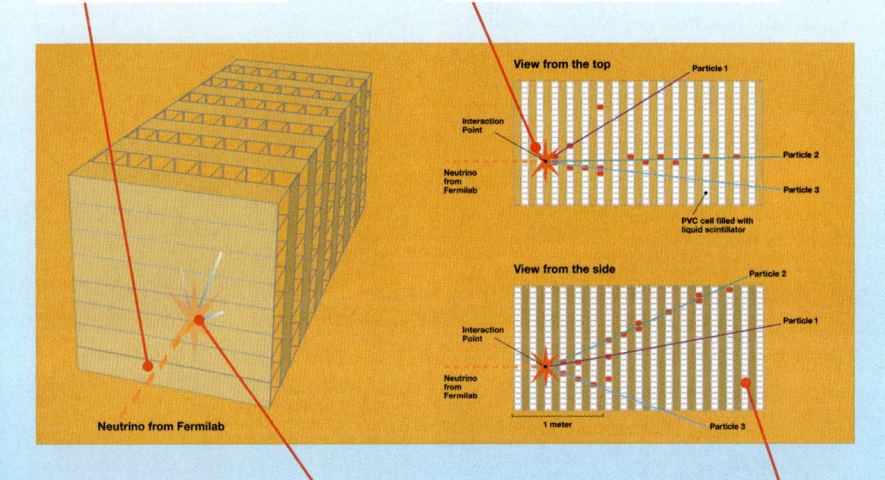

Noting the data
The data is picked up by electronics, which is then studied by physicists who will try to determine what the results mean.

Light me up
The scintillator in the detector lights up when it records a collision.

> "DUNE will be the largest international science experiment in the US"

Stay cool
The equipment is kept cool by liquid argon at -185 degrees Celsius to maintain superconduction.

Start of the journey
The particle accelerator below Fermilab creates a particle beam that is fired a distance of 1,300km.

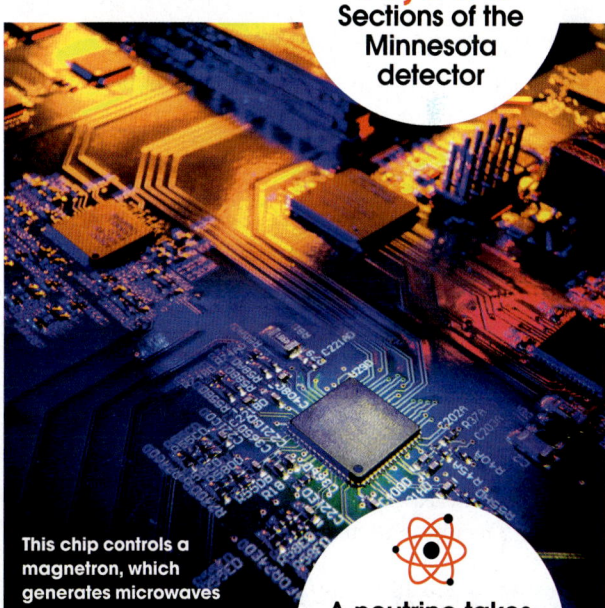

Learn more
WATCH NEUTRINO COLLISIONS LIVE
Visit nusoft.fnal.gov/nova/public to see live displays of the particle collisions, as recorded by the detectors at Ash River and Fermilab, a 360-degree video of the detector and more videos and graphics.

A Fermilab staff member working on the NOvA near detector

11,000
Sections of the Minnesota detector

This chip controls a magnetron, which generates microwaves

A neutrino takes just 0.0027 seconds to travel between the sites

Liquid argon at -185 degrees Celsius, that fills a cryostat that's used to test DUNE

© Fermilab; US Department of Energy

077

UNDERSTANDING PHYSICS

Extremes

80 Quantum power
The future of computing and how it will change your world

88 Extreme temperatures
Why leaving our usual range of temperatures makes materials do strange things

92 Deadly radiation
From dental X-rays to nuclear reactors: all you need to know about ionising radiation, its uses and hazards

100 The hidden universe
Dark matter and dark energy make up most of the universe, yet we can't see it. What is this strange stuff?

104 The power of atoms
Atoms are the ultimate construction kit, building everything from Venus de Milo to Venus the planet

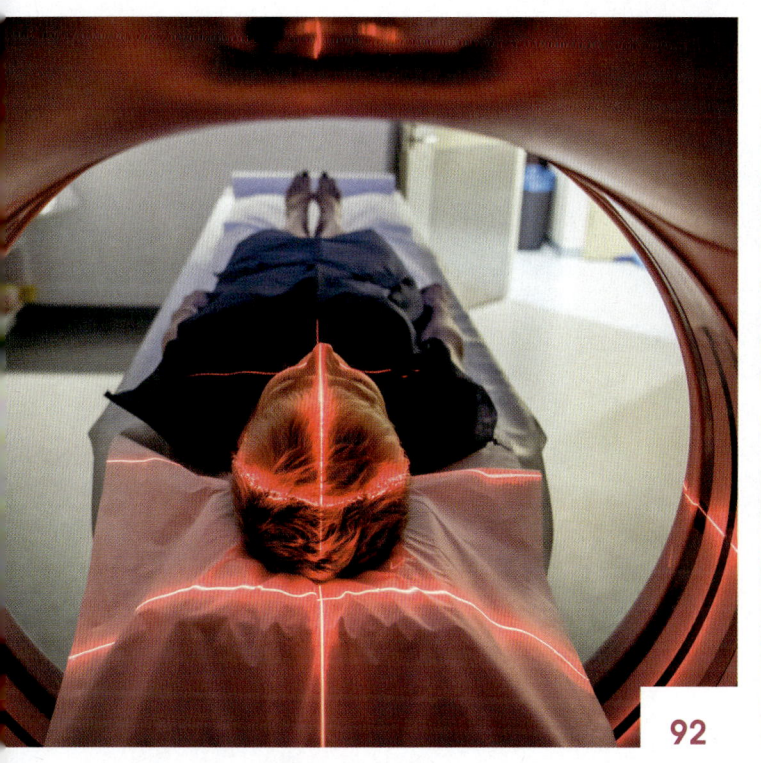

EXTREMES

92

100

104

79

UNDERSTANDING PHYSICS

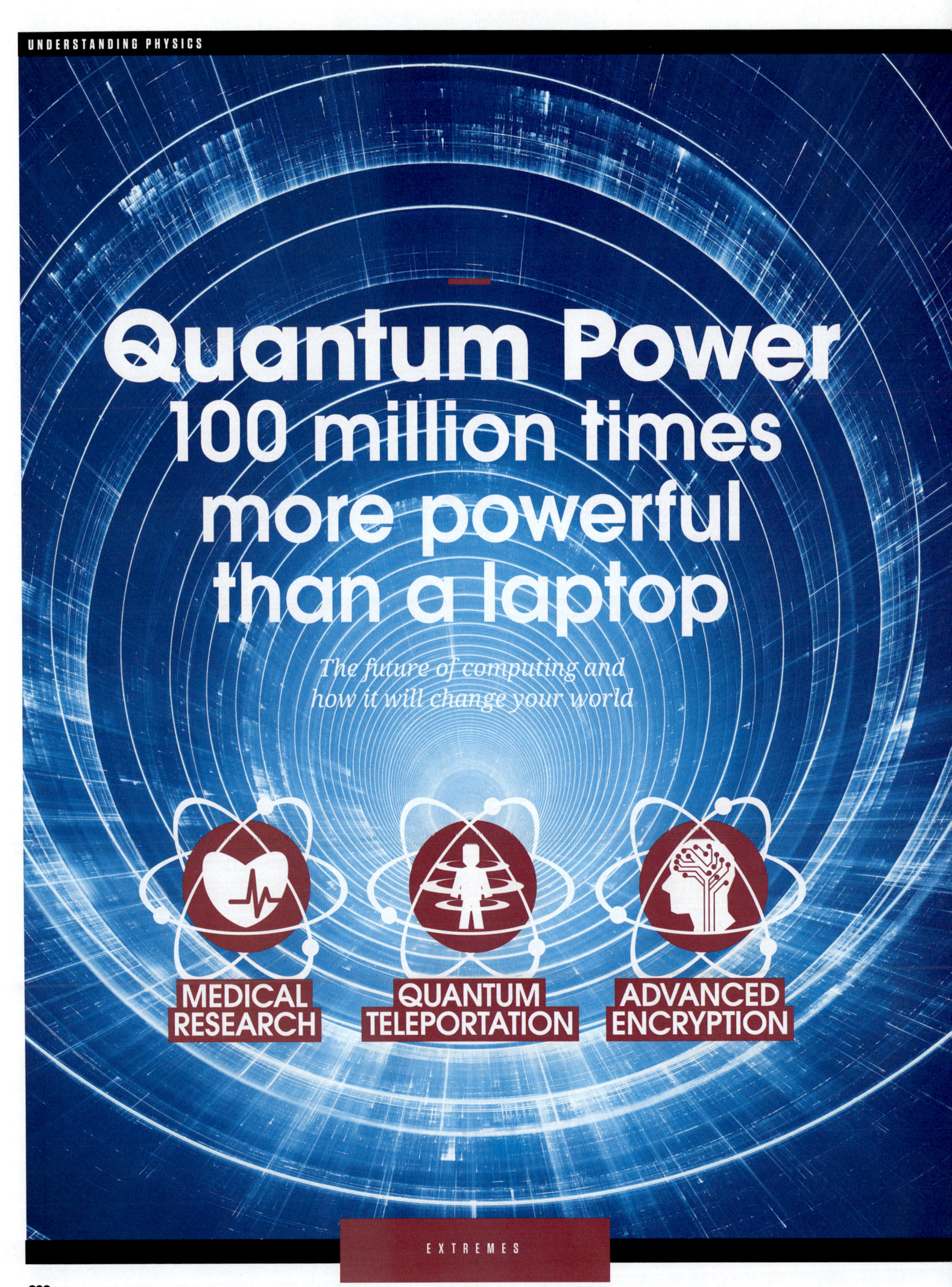

Quantum Power
100 million times more powerful than a laptop

The future of computing and how it will change your world

MEDICAL RESEARCH **QUANTUM TELEPORTATION** **ADVANCED ENCRYPTION**

EXTREMES

The pioneers of quantum mechanics

Introducing the people who dared to think the unthinkable, laying the foundations of quantum technology

Albert Einstein 1905
Einstein explained the photoelectric effect by suggesting that light took the form of discrete bundles called photons. This seemed at odds with light's wave nature.

Louis de Broglie 1923
French physicist Louis de Broglie expanded on previous discoveries by proposing that all tiny particles can behave as waves, and vice versa.

Erwin Schrödinger 1926
Austrian physicist Erwin Schrödinger's paper describing the motion of an electron as a wave function was a defining moment in quantum mechanics.

Werner Heisenberg 1925-1927
Alongside Niels Bohr, Werner Heisenberg suggested that subatomic particles only adopt a particular state when observed.

Alexander Holevo 1973
Russian mathematician Alexander Holevo was one of several researchers to lay down the theoretical foundations of quantum mechanics.

QUANTUM POWER

It might be a term that trips off the tongue, and it may suggest a field of study dominated by the scientific elite, but quantum mechanics – or quantum physics if you prefer – is largely a mystery to the layperson. Surprisingly, therefore, it couldn't be much simpler to sum it up, even though understanding it is considerably more difficult.

Quantum mechanics is concerned with the behaviour of atoms, photons and the various subatomic particles, and it contrasts with classical physics, which describes the behaviour of everyday objects that are large enough to see.

The difference between classical physics and quantum mechanics is absolutely staggering. The objects that we see in the world around us behave in a way that seems intuitive, but once we start to consider very small objects, intuition and common sense have to be abandoned.

Instead, when we consider them individually, atoms, electrons and photons behave in a way that most people would be inclined to describe as impossible. That perception of impossibility isn't a naive view either. Even the eminent Nobel Prize-winning physicist Niels Bohr is on record as saying "If anybody says he can think about quantum theory without getting giddy, it merely shows that he hasn't understood the first thing about it."

We'll look at some of these concepts in more detail in the boxout below, but, having made such an astonishing claim, it's surely only appropriate to provide a couple of examples of apparently impossible quantum behaviour.

Perhaps one of the most bizarre things that

> "The difference between classical physics and quantum mechanics is absolutely staggering"

Quantum concepts
Examining the bizarre quantum effects that underpin quantum technology

Superposition
A particle in superposition is in two states at once, so it could represent both a binary 0 and 1. Think of a coin: if it's spinning you can see heads and tails simultaneously.

Classical physics: Heads or tails
Quantum physics: Heads AND tails

Entanglement
Two entangled particles are strangely linked, so the fate of one affects the other. If you observe one particle this will cause its superposition to be lost, and the same will happen to its entangled twin.

Quantum physics:
Heads + Heads
Heads + Tails
Tails + Heads
Tails + Tails

N quantum bits or qubits — 2n possible states

Observation
Observing a particle in superposition causes it to adopt a single state. Any interaction with the environment does the same. The more entangled the particles, the harder it is to maintain superposition.

Observation or noise

No cloning
Making a copy of a particle in superposition also causes the superposition to be lost. This makes designing a quantum computer tricky, but, in quantum communications, it alerts the sender to the presence of an eavesdropper.

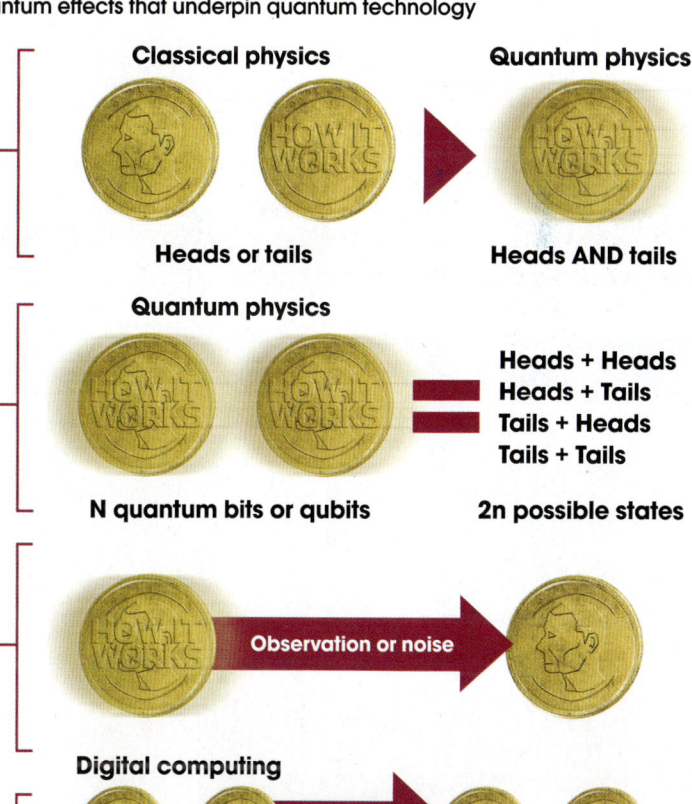

Digital computing — Copy or eavesdrop
Quantum computing — Copy or eavesdrop

UNDERSTANDING PHYSICS

Understanding quantum entanglement

Experiments have confirmed a quantum effect that Einstein called "spooky"

Crystals doped with the rare element neodymium could potentially store quantum memories

Splitting the beam
In this experiment a beam-splitter is used so that the two photons are dispatched to different destinations.

Generating entangled photons
By firing a laser beam through certain types of crystal, pairs of entangled protons can be created.

Superposition
Photon one is in a state of superposition, which means it's polarised horizontally and vertically at the same time.

Entanglement
Photon two is entangled with photon one, so they have a fixed relationship. They have the same or the opposite polarisation when they are in superposition.

Action at a distance
Observing one photon affects its entangled twin instantaneously, no matter how far apart the two photons are.

Effect on photon two
Because they're entangled, observing photon one will also have an effect on photon two, thereby fixing its polarisation.

A scientist at the University of Geneva in Switzerland uses a laser to create entangled photons in researching quantum memory

Observing photon one
When photon one is observed, superposition is lost and it will appear to be either horizontally (H) or vertically (V) polarised.

EXTREMES

082

QUANTUM POWER

Manipulating particles
The phenomenon that unlocks teleportation

Polarised photons
When it passes through a polarising filter light can become horizontally, vertically or diagonally polarised, which means that the photons spin only in one direction.

Unpolarised photons
Normal light is unpolarised, so each photon spins in all possible directions at the same time.

POLARISING FILTER

Defining photons
Moving through the filter dictates the state of the spin.

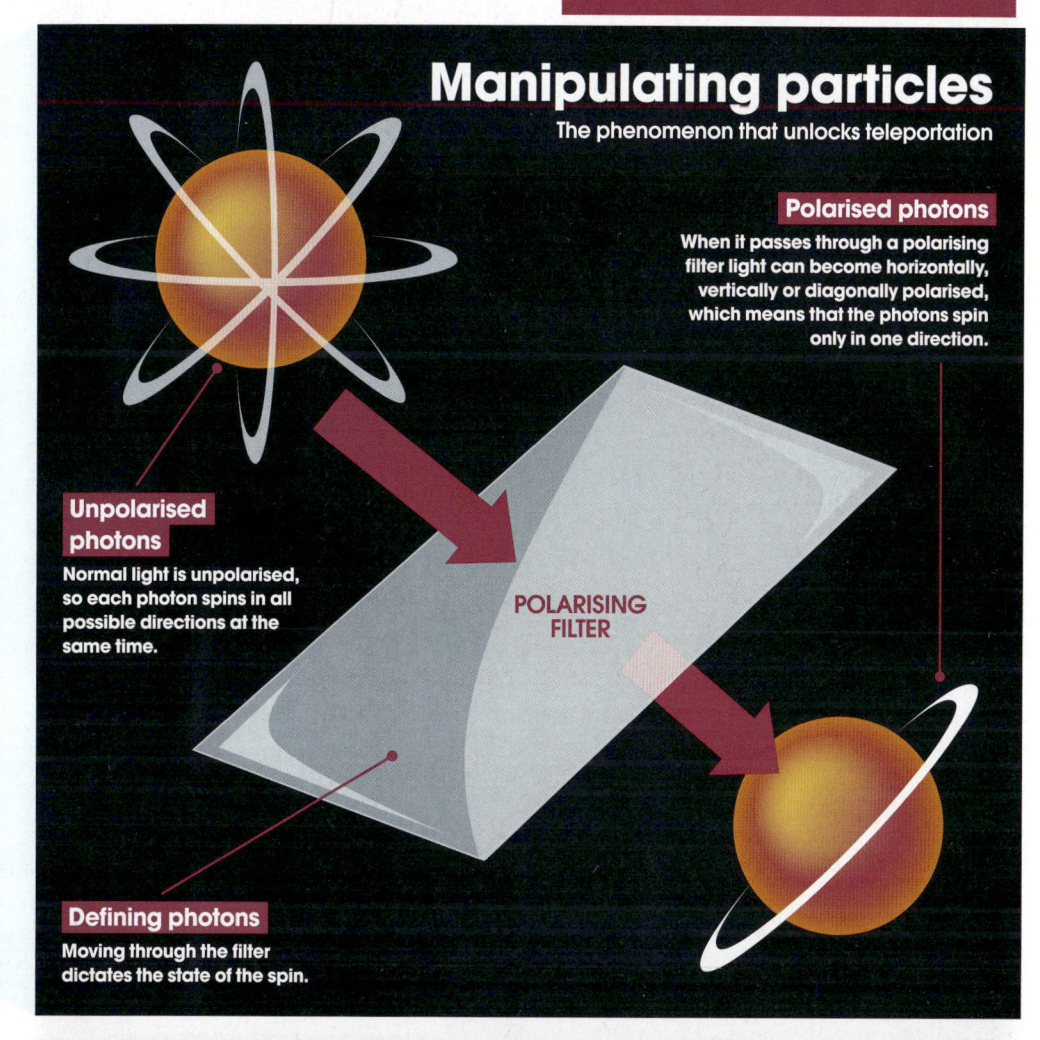

Quantum teleportation

We might be a considerable way from teleporting people, but single atoms and photons have already been teleported thanks to the use of quantum techniques.

The process involves creating two entangled particles at place A and then sending one of them to place B. Now, using clever techniques that also involve introducing a third particle that interacts with particle A, entanglement causes the particle at B to become an exact copy of the third particle. In reality, the actual particle hasn't moved, but the result is the same, so, effectively, the third particle has teleported to B's location instantly.

As with all things quantum, in practice this is incredibly difficult, and scientists are in a race to beat distance records. While the current record is 143 kilometres using a laser beam in September 2016, researchers in Calgary, Canada, and Shanghai, China, demonstrated quantum teleportation using a more practical fibre optic network to teleport photons across their respective cities.

Atoms and photons can now be teleported over ever-increasing distances

can happen in the subatomic realm is that objects such as electrons or photons can be in two places at the same time or in two different states at once – a so-called state of superposition. You'll never be able to observe this odd state of affairs because, as soon as you try to do so, that object will appear to be in just the one place or one state. However, scientists have conducted cunningly-devised experiments that have confirmed that this peculiar behaviour really does happen, despite indications to the contrary whenever we try to observe it.

Another strange effect is called quantum mechanical tunnelling, and it refers to the fact that a tiny object is capable of passing straight through a solid barrier without damaging it. So, for example, if you fire an electron at a sheet of gold foil, there's a possibility that it could appear at the other side with the foil still intact.

The fact that particles can be in two places at once, and that they can pass through solid objects, both stem from the dual nature of tiny objects. At one time light was thought of as a wave, but later it was discovered that it could be described as a stream of particles called photons. Conversely, electrons were once considered as miniature particles that orbited an atom's nucleus like planets orbiting a sun, but subsequently it

"Scientists have now taken their first steps in quantum teleportation"

was discovered that they could be described as wave functions.

In reality, electrons and photons each have the properties of both particles and waves, or, in other words, both concepts are correct. So, that strange phenomenon in which an electron can be in two places at once is a consequence of the wave nature of electrons.

Instead of that now outdated view of orbiting electrons, wave theory concerns a so-called probability function. In other words, it describes the probability of the electron being at any particular point in space and, until the electron is observed, its position can be thought of as all points in space, albeit with some places being more likely than others.

What we've seen so far has been known since the early 20th century, and it's strange enough. So any hope of ordinary people understanding the more recent developments in quantum theory is a forlorn one. However, to illustrate just how bizarre current thinking can be, let's think briefly about the multiverse theory, although even this dates back to the 1960s and 1970s.

UNDERSTANDING PHYSICS

You'll recall that observing a particle in a state of superposition causes its previously unknown position or state to become fixed. In the science-fiction-sounding multiverse theory, as soon as that observation takes place, the universe splits into two or more parallel universes, with that particle being in a different location or state in each version of reality. What's more, with vast numbers of these splits taking place each second, that soon gives us an unimaginable number of parallel universes. This theory has gained additional credence recently as scientists have started to consider quantum computers. As we'll see later here, compared to today's devices, if large-scale quantum computers ever become a reality, the performance they will offer will be absolutely astonishing.

This has led some scientists to suggest that there isn't enough material in the observable universe to carry out such a phenomenal number of computations. In the multiverse theory, however, that work is effectively farmed out into all of those parallel universes.

Given its very theoretical foundations, some people might be excused for thinking that quantum mechanics is an entertaining curiosity for scientists, but of absolutely no practical use. But experience has proven that most theoretical studies impact the real world eventually, and there's every indication that the same is true of studies in the quantum realm. Quantum mechanics has already given birth to many technological breakthroughs, and there are

The D-Wave 2X quantum computer
The secrets of one Canadian company's greatest creation

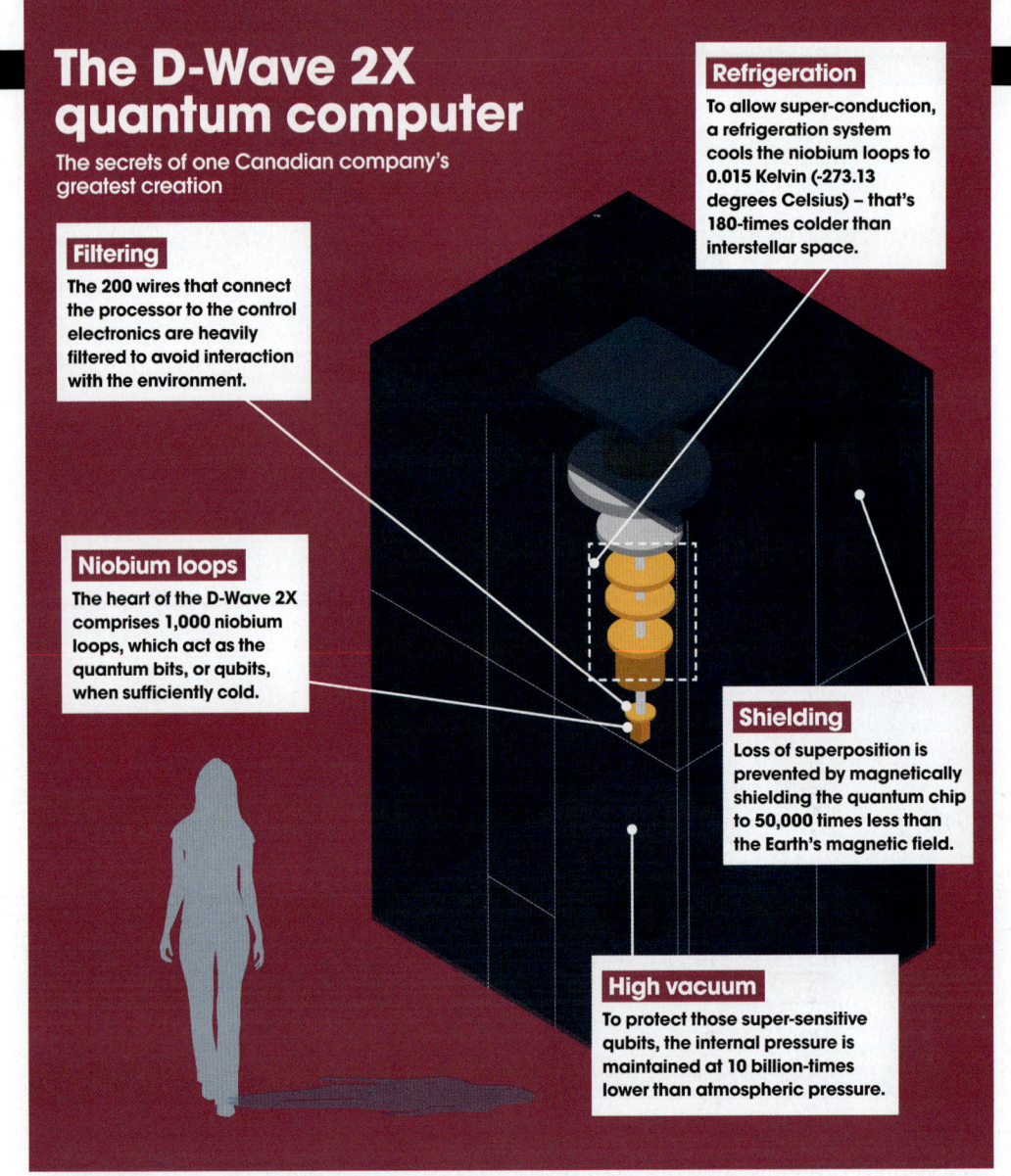

Filtering
The 200 wires that connect the processor to the control electronics are heavily filtered to avoid interaction with the environment.

Niobium loops
The heart of the D-Wave 2X comprises 1,000 niobium loops, which act as the quantum bits, or qubits, when sufficiently cold.

Refrigeration
To allow super-conduction, a refrigeration system cools the niobium loops to 0.015 Kelvin (-273.13 degrees Celsius) – that's 180-times colder than interstellar space.

Shielding
Loss of superposition is prevented by magnetically shielding the quantum chip to 50,000 times less than the Earth's magnetic field.

High vacuum
To protect those super-sensitive qubits, the internal pressure is maintained at 10 billion-times lower than atmospheric pressure.

1 Quantum annealer
Today's only commercial quantum computer is a quantum annealer. This is a specialised architecture that is designed for a whole range of applications that are described as optimisation tasks.
DIFFICULTY LEVEL

2 Analogue quantum
Before digital computers were fast enough, high-speed scientific calculations were performed using analogue computers. In the same way, quantum analogue computers could provide an interim solution until universal machines appear.
DIFFICULTY LEVEL

3 Universal quantum
Like today's computers, a universal quantum computer would be able to perform any type of computation, but would be almost immeasurably faster thanks to superposition and entanglement.
DIFFICULTY LEVEL

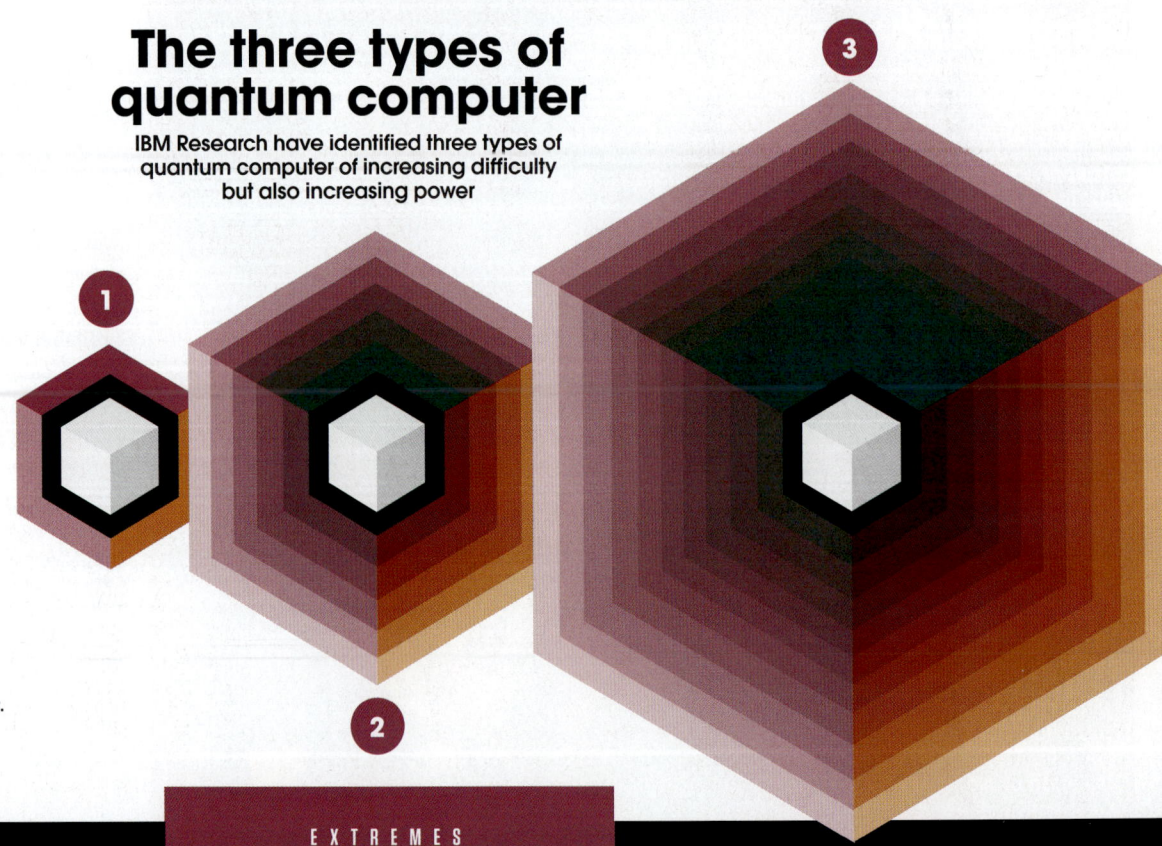

The three types of quantum computer
IBM Research have identified three types of quantum computer of increasing difficulty but also increasing power

EXTREMES

084

QUANTUM POWER

Qubits – the secret of quantum computing

This peculiar quantum effect is key to quantum computing and several other quantum technologies

Binary arithmetic
Conventional digital computers operate on binary arithmetic in which all numbers are a sequence of bits, either 0s or 1s.

Up and down arrows
Another way of looking at the 0s and 1s of traditional computers is as arrows – say up for 1 and down for 0.

The quantum equivalent
In quantum computers, bits are called qubits (quantum bits) and they are represented by single tiny particles.

The globe analogy
The state of a qubit can be represented as an arrow from the centre to a point on the circumference of a sphere.

Electrical currents
In ordinary computers, 0s and 1s are represented by an electrical current or, in other words, the effect of lots of electrons.

Superposition
Arrows to other points on the sphere's circumference represent superpositions – varying degrees of 1 and 0 simultaneously.

1s and 0s
As with ordinary bits, arrows to the north and south pole represent 1s and 0s.

Measurement
When you read a qubit its value will always be 0 or 1, the probability of each depending on its latitude. This makes it tricky to devise algorithms that can capitalise on the potential of quantum computation.

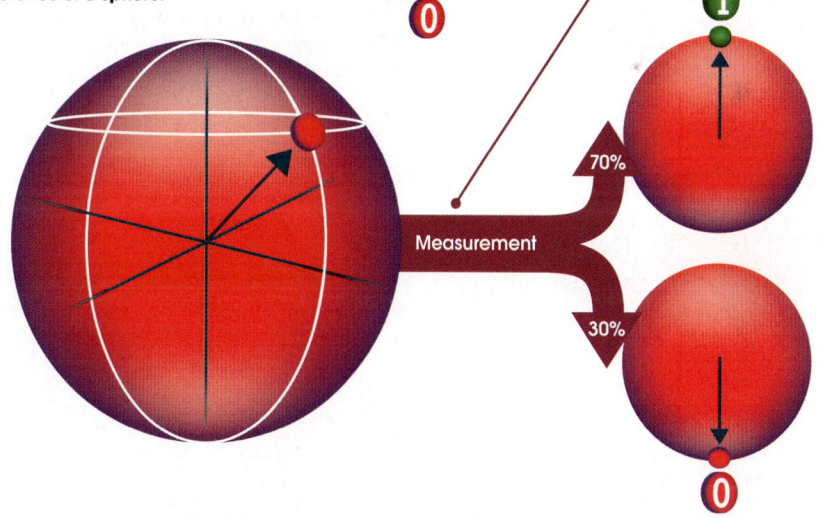

MRI scanners work using the principles of quantum mechanics

tantalising glimpses of what may lie ahead.

For a start, today's solid-state devices, which impact so much of 21st century life, depend on quantum effects. Most importantly, perhaps, is the transistor, which is the fundamental building block of computers, smartphones and pretty much all electronic devices. Another important solid-state device is the LED and the closely related solid-state laser. The former is now revolutionising lighting by bringing hitherto unprecedented levels of energy efficiency, while the latter is key to the fibre optic cables that span the globe empowering the internet, and is also a vital component in CD and DVD drives.

Atomic clocks are also reliant on quantum mechanics, and these instruments provide the precision timing needed for the operation of GPS systems on which sat navs and smartphone navigation apps rely. Quantum mechanics also underlies the principles of magnetic resonance imaging (MRI) machines, which allow physicians to see inside the body.

Little was said about their quantum heritage when these various technologies were developed, but we're now starting to hear about several new technologies that are much more up front about their quantum roots. What's more, these up-and-coming applications of quantum mechanics are absolutely mind-blowing.

Quantum computers by numbers

100 million times
How much faster the D-Wave 2X is compared to an ordinary computer

$2^{1,000}$
Number of solutions the current D-Wave 2X quantum computer can search simultaneously

1,000
The greatest number of entangled qubits achieved

18.4 billion billion
How many calculations a universal 64-qubit quantum computer could do simultaneously

2^{16}
How many simultaneous searches the first 16-qubit D-Wave quantum computer could perform in 2007

100,000
Number of qubits needed for a practical universal quantum computer

085

UNDERSTANDING PHYSICS

Thought that Star Trek style teleportation was the result of an over-active imagination? Think again – scientists have now taken their first steps in quantum teleportation. What about a code that is totally unbreakable? Experience tells us that however sophisticated a code, all it takes is a sufficiently powerful computer and encrypted messages can be accessed. Not so with quantum cryptography. This isn't a code that's so fiendishly difficult that it would take all the computers on the planet years to crack. This is a method of encryption that, according to the laws of quantum mechanics, is totally secure, however much computing power you throw at it. And then we have quantum computers and the world of opportunities it opens up.

For now, though, let's just say that one company is already selling a rather specialised quantum computer, and research continues into a quantum equivalent of today's PCs, a universal quantum computer. If these ever come to fruition, they won't just be incrementally faster than their predecessors, which have doubled in speed every couple of years or so. Instead, a truly universal quantum computer holds the promise of almost unlimited computing performance thanks to that strange quantum effect of superposition, coupled with the equally strange quantum effect of entanglement.

By being in millions upon billions of states at the same time, a universal quantum computer would offer the ultimate in massively parallel processing, in which multiple operations are carried out simultaneously.

It is widely acknowledged that last century was the era of electronics. Within a period of just 52 years the very first electronic device, the vacuum tube or valve, was invented, and this was superseded first by the transistor and then by the integrated circuit. It only took another 13 years for the first microprocessor to be released.

Renowned quantum physicist Professor Rainer Blatt has described the technological developments of the last century as the first quantum revolution, and with some justification. After all, many of the developments that underpin today's society resulted from an understanding of quantum mechanics and, in particular, wave-particle duality. Professor Blatt suggests that humanity now stands at the dawn of a second quantum revolution that will be empowered by the weird quantum effect of entanglement.

According to Professor Blatt, "In the early 1960s, the laser was still seen as a solution to an unknown problem, and today, just over 50 years later, lasers have become an indispensable part of our lives – I expect quantum technologies to develop along similar lines."

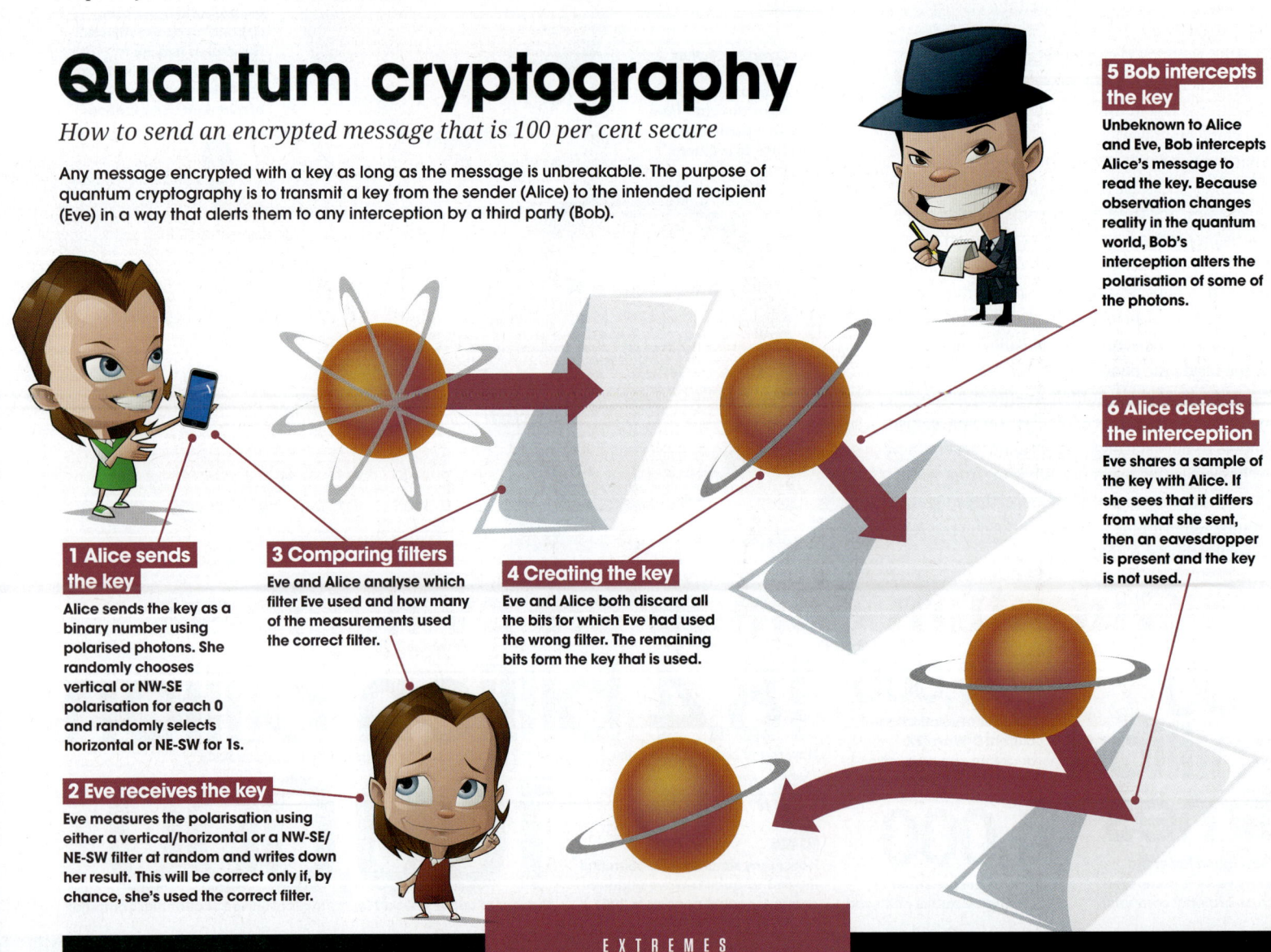

Quantum computing has a wide range of applications, from improving air-traffic control systems to creating better speech recognition software

Quantum cryptography
How to send an encrypted message that is 100 per cent secure

Any message encrypted with a key as long as the message is unbreakable. The purpose of quantum cryptography is to transmit a key from the sender (Alice) to the intended recipient (Eve) in a way that alerts them to any interception by a third party (Bob).

1 Alice sends the key
Alice sends the key as a binary number using polarised photons. She randomly chooses vertical or NW-SE polarisation for each 0 and randomly selects horizontal or NE-SW for 1s.

2 Eve receives the key
Eve measures the polarisation using either a vertical/horizontal or a NW-SE/NE-SW filter at random and writes down her result. This will be correct only if, by chance, she's used the correct filter.

3 Comparing filters
Eve and Alice analyse which filter Eve used and how many of the measurements used the correct filter.

4 Creating the key
Eve and Alice both discard all the bits for which Eve had used the wrong filter. The remaining bits form the key that is used.

5 Bob intercepts the key
Unbeknown to Alice and Eve, Bob intercepts Alice's message to read the key. Because observation changes reality in the quantum world, Bob's interception alters the polarisation of some of the photons.

6 Alice detects the interception
Eve shares a sample of the key with Alice. If she sees that it differs from what she sent, then an eavesdropper is present and the key is not used.

EXTREMES

QUANTUM POWER

Applying quantum mechanics

In the future there may be numerous ways to use this rapidly growing technology

Image searching

Humans are easily able to detect familiar objects such as trees, lakes and cats when they look at a photograph. Teaching computers to do likewise is a really difficult programming task because it's so hard to define the essence of 'cat-ness', for example. But machine learning tasks like this are natural applications for quantum computers. Google has already invested considerable research effort into image analysis to make online picture searching more efficient. This was demonstrated by teaching a quantum computer to recognise cars in photos, a task it was then able to do much quicker than an ordinary computer could ever achieve.

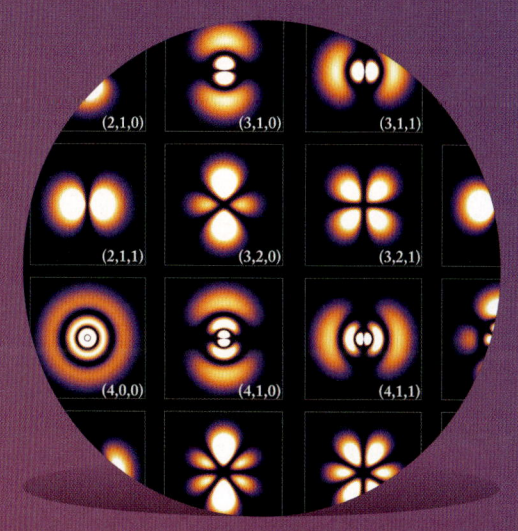

Quantum simulation

It might seem like a circular argument but, just as an understanding of quantum mechanics has now given us quantum computers, scientists are now hoping that those same quantum computers will help them to better understand quantum systems by simulating them. Today's computers are able to carry out simulations of quantum effects but, such is the complexity of quantum systems, they are incredibly slow. It perhaps comes as no surprise, then, that computers that are based on the strange world of quantum mechanics are much more capable of simulating quantum systems and, thereby, help scientists to gain new insights.

Astronomy

Given that NASA jointly owns one of the world's first quantum computers, it's hardly surprising that astronomy will probably be one of the main beneficiaries of this new model of computation. The space agency has its sights set on several ways that quantum computing can assist in the exploration of space, but many of them can be summed up as searching through huge amounts of data for the proverbial needle in the haystack. A classic example is the search for habitable exoplanets; Earth-like planets in orbit at the ideal distance from faraway stars, that might just be capable of hosting life.

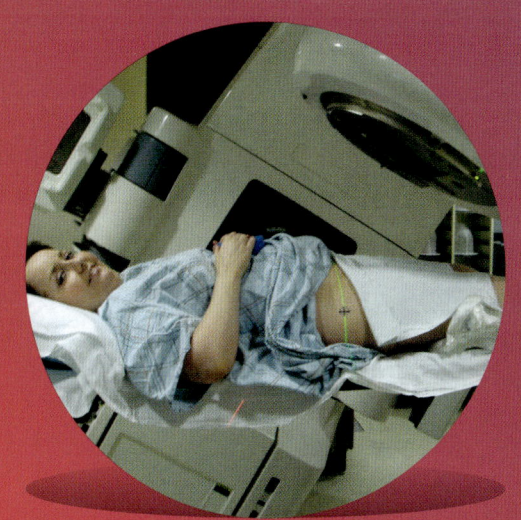

Radiotherapy optimisation

According to D-Wave Systems, their D-Wave 2X computer, working with a conventional computer, will help to optimise radiotherapy. This treatment aims to target a tumour while minimising harmful exposure to the rest of the body, with several beams intersecting at the tumour. Its optimisation involves juggling thousands of variables. To achieve it, simulations would be carried out on a huge number of possibilities using a conventional computer, while a quantum annealing computer would determine the most probable scenarios for simulation.

Code breaking

A universal quantum computer would be able to factor large numbers with ease, a phenomenally time-consuming job for conventional computers. Today's ciphers rely on the fact that factoring is difficult, but encrypted messages would be an open book once general-purpose quantum computers become reality. This might be useful to the military and police forces – for example, in the fight against terror and organised crime – but it would also be a boost to cyber criminals. It's quite appropriate that the same quantum technology that might make today's encryption techniques obsolete could provide a replacement in the form of quantum cryptography.

"Encrypted messages would be an open book for a general-purpose quantum computer"

UNDERSTANDING PHYSICS

Extreme temperatures

Why leaving our usual range of warm and cold makes materials do strange and exciting things

EXTREMES

EXTREME TEMPERATURES

You might be surprised by how mind-bending the idea of temperature can be, even though it's such an obvious part of our everyday lives. Few things are as easy to imagine as the shivering cold in winter and the warming sunlight in summer. But we live within a narrow range of all the temperatures that are possible. Outside of this Goldilocks window lie hot and cold extremes where everything becomes very strange. Go cold enough and the air around us would turn solid, turning everything into a single giant brick. Go hot enough and atoms no longer hold together, creating weird states of matter that we could not survive an encounter with.

Even temperature itself is a strange idea. We use the word to talk about how energy affects matter, the physical stuff that makes up the universe. Temperature rises when energy flows into matter, and falls when it leaves. In our everyday lives, as an object's temperature rises the atoms and molecules inside it jiggle around more. As its temperature cools, atoms and molecules in the object jiggle more slowly, becoming stiller. You can show this by dropping food colouring into water. The hotter the water is, the faster the colouring will spread, because water molecules are jiggling around and jostling it more.

Scientists care so much about this they've invented several temperature scales to measure it. Two common scales break up temperature into degrees Celsius or Kelvin. These scales are similar, with the difference between one and two degrees Celsius the same as the difference between one and two Kelvin. The point at which water molecules slow down enough to go from liquid to solid is zero degrees Celsius, but that corresponds to 273.15 Kelvin. That is because the Kelvin scale starts at what is supposed to be the lowest temperature anything can reach – zero Kelvin, which is -273.15 degrees Celsius. Here all atoms would be perfectly still, but physics gets weird and messes up that idea. Many scientists think that nothing can actually reach zero Kelvin, but others have suggested that negative temperatures on the Kelvin scale might be possible. Either way, a colder world would be very different long before that. Most of the air we breathe is made up of nitrogen gas, which turns liquid at 77 Kelvin and solid at 63 Kelvin.

In the other direction, it's easy to imagine that heating matter up makes its atoms and molecules move ever faster. But at temperatures of thousands of Kelvin, like you find on the surface of stars, the heat will make molecules fall apart into the atoms they're made of. If you continue to turn up the heat, the atoms themselves will fall apart, leading to ionised plasma. And scientists think that you can, in principle, go much higher. Theories say that at 2 billion Kelvin and beyond, the nucleus at every atom's heart will fall apart. The upper temperature limits are very high indeed, imposed by the universe running out of energy, or changing its nature altogether. Such hot prospects are truly awesome to imagine.

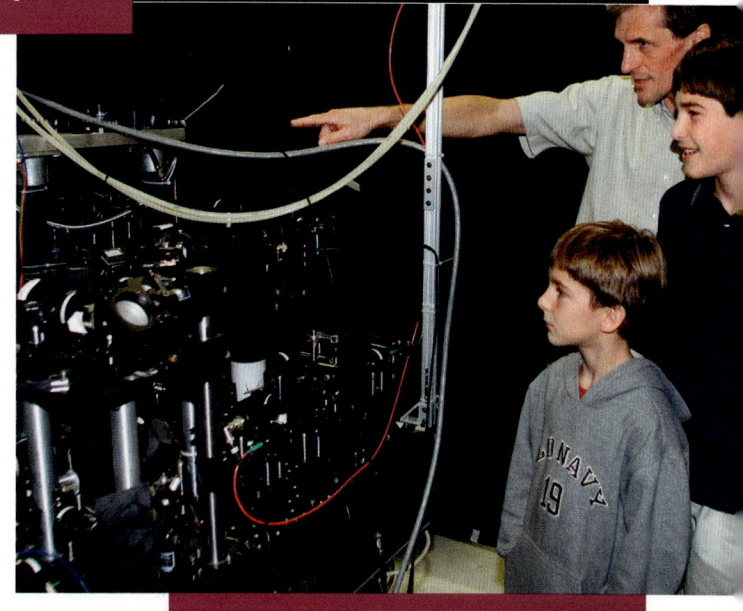

Dr Wolfgang Ketterle, the scientist that holds the record for cooling atoms closest to absolute zero with his lasers

Absolutely impossible zero

According to traditional science, absolute zero is the coldest possible temperature. When atoms and molecules are near absolute zero, they should stop moving altogether. But scientists cannot fully achieve this state of zero heat energy in a substance. That's because at this extreme of temperature, physics gets very strange. Quantum physics laws, which govern how electrons behave, say that there must always be some energy to make molecules move. The same weird principles led scientists to suggest that negative temperatures on the Kelvin scale might be possible. This idea is controversial. Such negative Kelvin temperatures have never been measured, and scientists argue over whether or not they can really exist. Researchers have shown, however, that molecules cooled to near absolute zero can react chemically. That shouldn't be possible according to traditional science, but weird quantum physics effects allow it to happen.

Cold or hot stuff?
Changing matter's temperature affects its energy, pushing it through phase transitions between different states

1 Ionisation and recombination
Rare on Earth, plasma forms naturally in stars. There, high temperatures strip electrons from atoms, ionising them. At lower temperatures the electrons and the atoms' nuclei recombine into a gas.

2 Sublimation and deposition
Materials like solid carbon dioxide, or dry ice, sublime into a gassy cloud without going through a liquid form. The reverse process straight from gas to solid is called deposition.

3 Vaporisation and condensation
Steam rising from a kettle originates as liquid water vaporises. Condensation is the reverse phase change, such as water droplets forming from vapour on colder tiles next to a kettle.

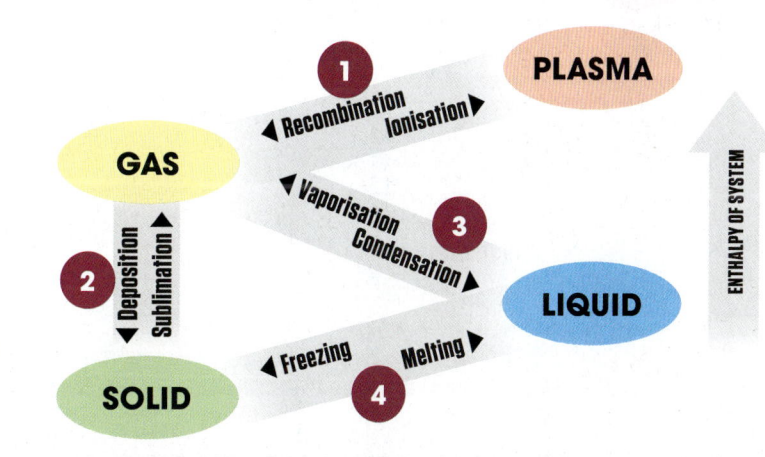

4 Melting and freezing
Changes from solid to liquid and back as substances melt and freeze are perhaps the most familiar phase transitions. Think of ice melting then turning solid again in the freezer.

UNDERSTANDING PHYSICS

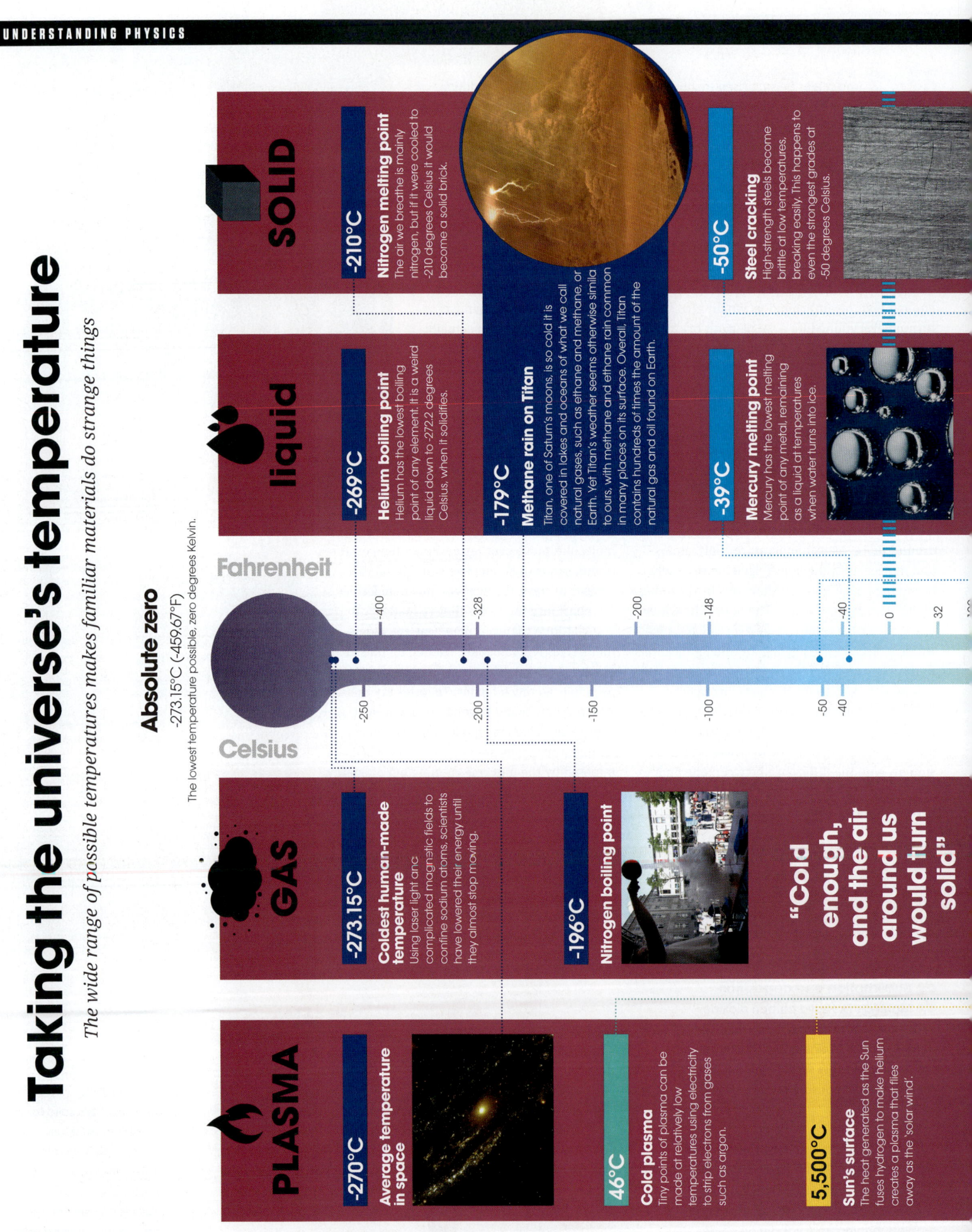

Taking the universe's temperature

The wide range of possible temperatures makes familiar materials do strange things

Absolute zero
-273.15°C (-459.67°F)
The lowest temperature possible, zero degrees Kelvin.

SOLID

-210°C
Nitrogen melting point
The air we breathe is mainly nitrogen, but if it were cooled to -210 degrees Celsius it would become a solid brick.

-50°C
Steel cracking
High-strength steels become brittle at low temperatures, breaking easily. This happens to even the strongest grades at -50 degrees Celsius.

liquid

-269°C
Helium boiling point
Helium has the lowest boiling point of any element. It is a weird liquid down to -272.2 degrees Celsius, when it solidifies.

-179°C
Methane rain on Titan
Titan, one of Saturn's moons, is so cold it is covered in lakes and oceans of what we call natural gases, such as ethane and methane, or Earth. Yet Titan's weather seems otherwise similar to ours, with methane and ethane rain common in many places on its surface. Overall, Titan contains hundreds of times the amount of the natural gas and oil found on Earth.

-39°C
Mercury melting point
Mercury has the lowest melting point of any metal, remaining as a liquid at temperatures when water turns into ice.

Fahrenheit
-400
-328
-200
-148
-40
0
32

Celsius
-250
-200
-150
-100
-50
-40

GAS

-273.15°C
Coldest human-made temperature
Using laser light and complicated magnetic fields to confine sodium atoms, scientists have lowered their energy until they almost stop moving.

-196°C
Nitrogen boiling point

"Cold enough, and the air around us would turn solid"

PLASMA

-270°C
Average temperature in space

46°C
Cold plasma
Tiny points of plasma can be made at relatively low temperatures using electricity to strip electrons from gases such as argon.

5,500°C
Sun's surface
The heat generated as the Sun fuses hydrogen to make helium creates a plasma that flies away as the 'solar wind'.

EXTREMES

090

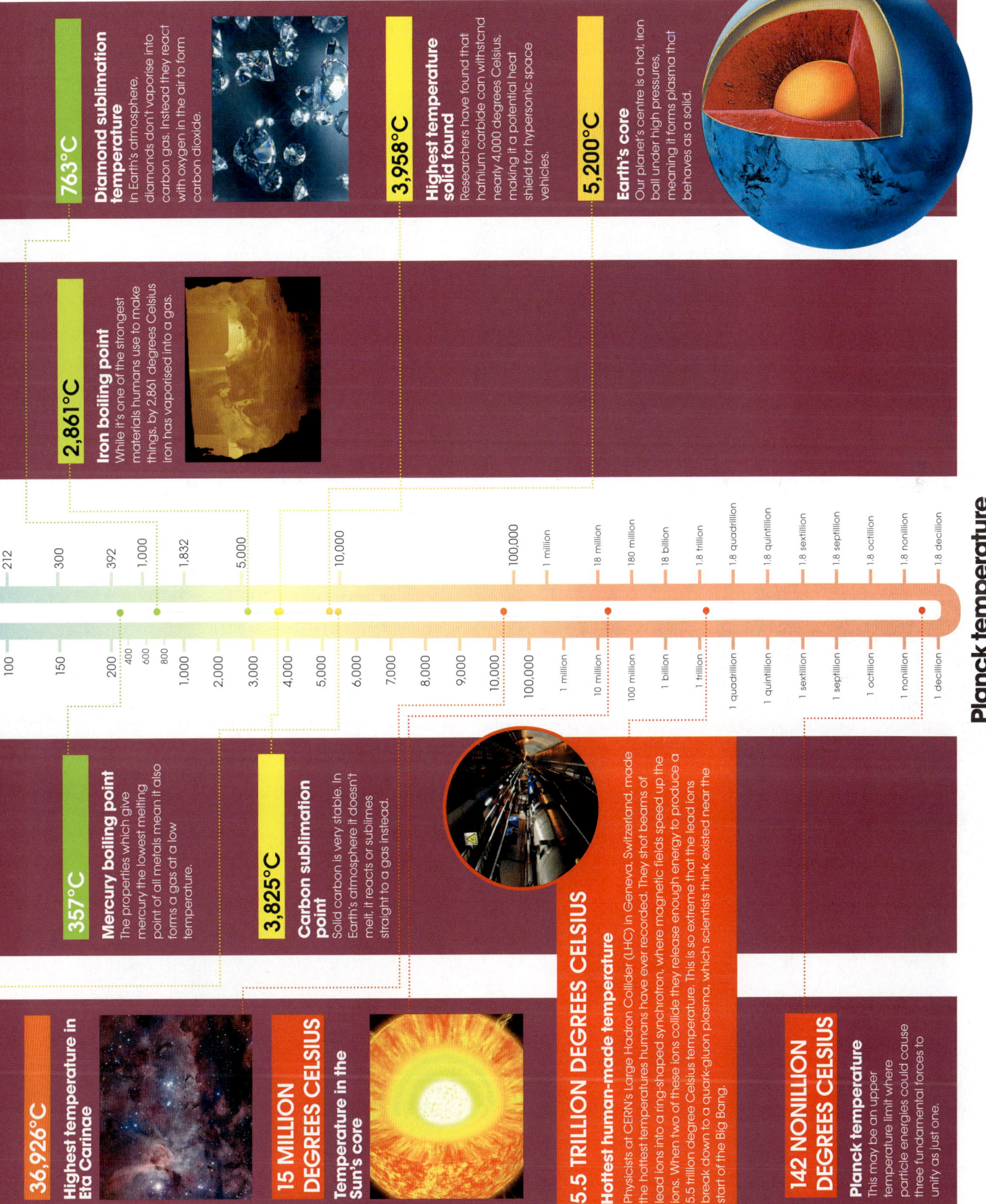

EXTREME TEMPERATURES

763°C
Diamond sublimation temperature
In Earth's atmosphere, diamonds don't vaporise into carbon gas. Instead they react with oxygen in the air to form carbon dioxide.

3,958°C
Highest temperature solid found
Researchers have found that hafnium carbide can withstand nearly 4,000 degrees Celsius, making it a potential heat shield for hypersonic space vehicles.

5,200°C
Earth's core
Our planet's centre is a hot iron ball under high pressures, meaning it forms plasma that behaves as a solid.

2,861°C
Iron boiling point
While it's one of the strongest materials humans use to make things, by 2,861 degrees Celsius iron has vaporised into a gas.

357°C
Mercury boiling point
The properties which give mercury the lowest melting point of all metals mean it also forms a gas at a low temperature.

3,825°C
Carbon sublimation point
Solid carbon is very stable. In Earth's atmosphere it doesn't melt. It reacts or sublimes straight to a gas instead.

36,926°C
Highest temperature in Eta Carinae

15 MILLION DEGREES CELSIUS
Temperature in the Sun's core

5.5 TRILLION DEGREES CELSIUS
Hottest human-made temperature
Physicists at CERN's Large Hadron Collider (LHC) in Geneva, Switzerland, made the hottest temperatures humans have ever recorded. They shot beams of lead ions into a ring-shaped synchrotron, where magnetic fields speed up the ions. When two of these ions collide they release enough energy to produce a 5.5 trillion degree Celsius temperature. This is so extreme that the lead ions break down to a quark-gluon plasma, which scientists think existed near the start of the Big Bang.

142 NONILLION DEGREES CELSIUS
Planck temperature
This may be an upper temperature limit where particle energies could cause three fundamental forces to unify as just one.

Planck temperature
142,000,000,000,000,000,000,000,000,000,000°C (255,000,000,000,000,000,000,000,000,000,000°F)
The hottest theoretical temperature of matter achievable

UNDERSTANDING PHYSICS

EXTREMES

DEADLY RADIATION

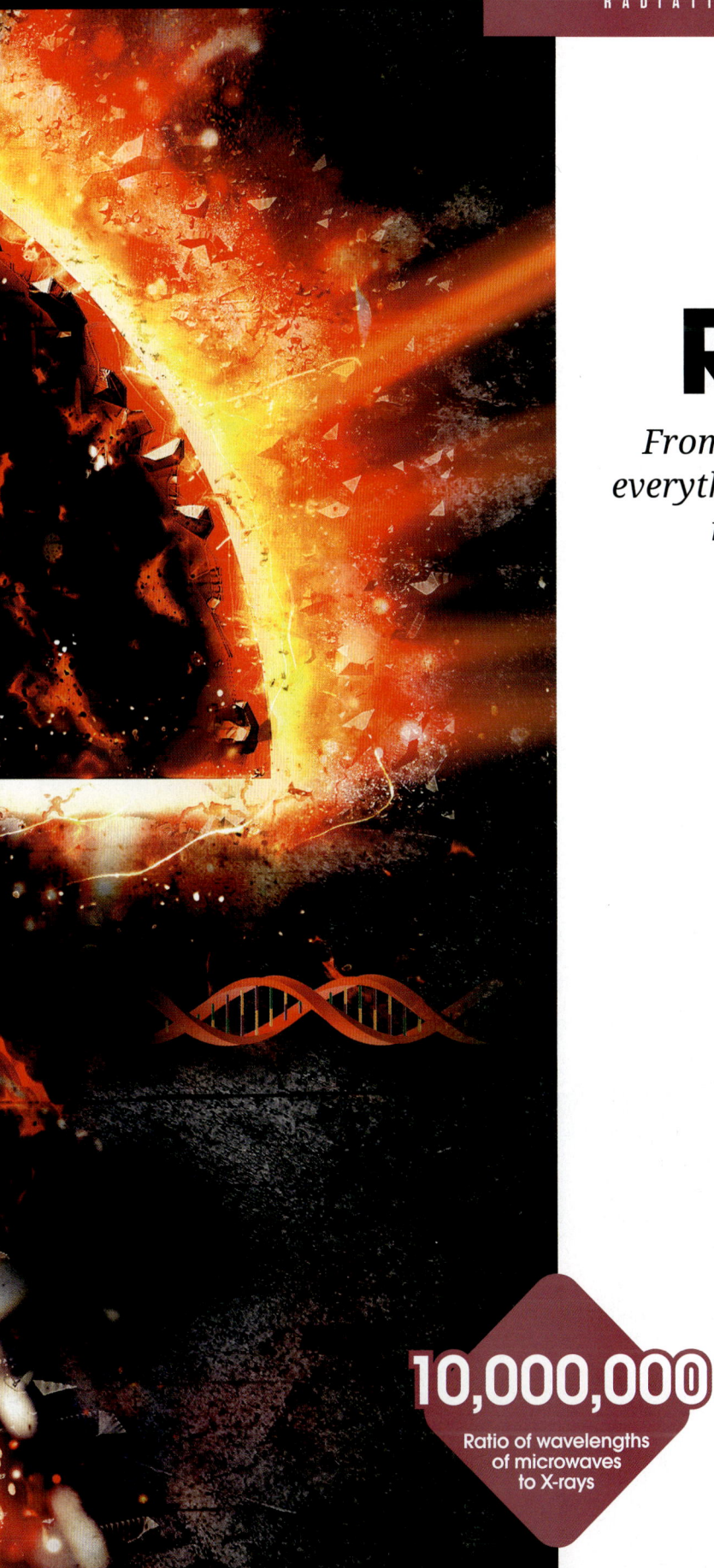

Deadly Radiation

From dental X-rays to nuclear reactors: everything you need to know about ionising radiation, its uses and hazards

For many, the word 'radiation' can set alarm bells ringing. If you think of the terrible death toll of the Hiroshima and Nagasaki bombs, or the devastating environmental effects of the Chernobyl nuclear disaster, it's easy to see why. Events like these show just how lethal radiation can be. But radiation is around us all the time – most of it completely harmless, and much of it positively beneficial.

In its broadest sense, radiation refers to any form of emitted energy that travels outwards – or radiates – from a source. It can take the form of streams of fast-moving particles, such as those emitted by radioactive materials, or of the electromagnetic (EM) waves generated when electrons jump from one energy level to another inside atoms. The heat and light we receive from the Sun come to us in the form of EM radiation, and life on Earth would be impossible without these particular forms of radiation.

There are other types of EM radiation that we can produce by technological means, such as radio for communication, microwaves for cooking or X-rays for medical imaging. Although these are created and used in different ways, they are all essentially the same type of wave, travelling from A to B at the same speed – the speed of light. The difference lies in wavelength and frequency. EM radiation has a spectrum ranging from radio waves at the low frequency, long-wavelength end to X-rays and gamma rays at the high frequency, short wavelength.

Any kind of EM radiation can be dangerous if enough energy is pumped into it. You see hazard warnings on microwave ovens, for example, and on lasers – which are simply an intense beam of light. But there's a subtlety which makes the

10,000,000
Ratio of wavelengths of microwaves to X-rays

UNDERSTANDING PHYSICS

Hans Geiger (left) with Ernest Rutherford, who named alpha, beta and gamma radiation

"The Earth is under constant bombardment from cosmic radiation"

high-frequency end of the EM spectrum substantially more dangerous than the lower end. This stems from the fact that EM radiation is fundamentally bipolar – the so-called 'wave-particle duality'. Although it travels from its source to its destination exactly as if it were a wave, when it gets there it passes on its energy as if it were packaged up in discrete particles, called photons. The higher the frequency, the more energy each photon carries. If you have a microwave beam and an X-ray beam with the same total energy, the microwave energy will be spread out over millions of times more photons.

The significance of discrete photons lies in the effect they have when they hit atoms at the receiving end. If there are a lot of low-energy photons, as in a microwave beam, they just cause the atoms to vibrate a bit more, which has the effect of increasing the temperature. The same is true of infrared radiation, which lies between microwaves and visible light in the spectrum, and is the principal way that the Sun heats the Earth. When you feel warm sunshine on your face, it's because the photons are causing the atoms to jiggle about just that bit more.

It's different on the other side of the visible spectrum, where we find higher frequency ultraviolet (UV) radiation. UV photons carry enough energy to produce internal changes in atomic structure, boosting electrons to higher energy levels and altering molecular bonds. This can potentially cause DNA damage, which is why we have to wear sunscreen when UV levels are high.

As we go further up the frequency spectrum, we get to a point where individual photons have so much energy they don't simply boost

400x Radioactivity from Chernobyl exceeded the Hiroshima bomb

One of dozens of infamous atomic weapons tests at Bikini Atoll, 1946 to 1958. A fleet of dummy warships is overwhelmed by the blast

That spooky blue glow

Although ionising radiation is invisible, we sometimes see its effects on surrounding material. Old-style radioluminescent paint glows because it's slightly radioactive, stimulating light emission from the paint molecules. More spectacularly, water-cooled nuclear reactors emit an eerie blue glow called Cherenkov radiation. This isn't dangerous itself – it's just ordinary light – but it's caused by super-fast beta particles that would be harmful if they weren't absorbed by the water, which has a protective as well as cooling function.

The beta particles travel close to the speed of light, but paradoxically the light itself doesn't. That's because light slows down to three-quarters of its normal speed when it travels through water. The result is a shock wave – like the sonic boom caused by an aircraft travelling faster than the speed of sound – and that's what we see as the characteristic Cherenkov glow.

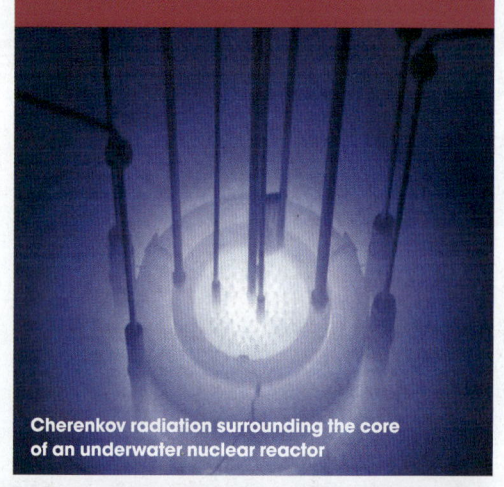

Cherenkov radiation surrounding the core of an underwater nuclear reactor

Nuclear tests 1945 to 1996
The nuclear tests carried out during the 20th century added to background radiation levels

45 CHINA

45 UNITED KINGDOM

210 FRANCE

715 SOVIET UNION

1,032 UNITED STATES

EXTREMES

DEADLY RADIATION

Radiation shielding
Some types of ionising radiation are easier to block than others

Alpha particles
Of the two types of particle radiation emitted by radioactive materials, alpha rays are the easiest to stop. A thin sheet of paper will do it – or, if you want to live dangerously, a human hand.

Gamma rays
The third type of radiation emitted by radioactive substances is even harder to block, because it consists of EM waves rather than particles. But even gamma rays can be stopped by a few centimetres of lead shielding.

Neutron radiation
The fast neutrons created in a nuclear reactor are the hardest type of radiation to stop. Heavy elements like lead aren't as effective as light ones like hydrogen, so reactors are shielded by several metres of hydrogen-rich concrete or water.

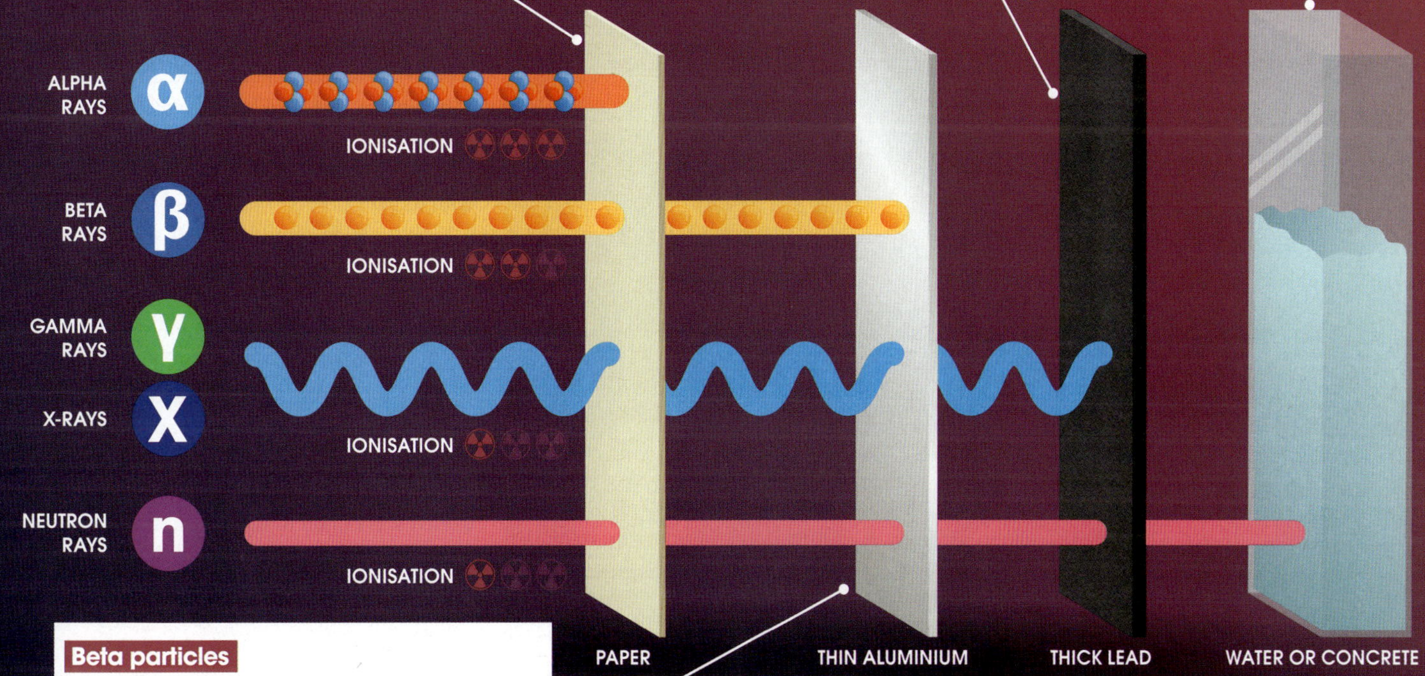

ALPHA RAYS α — IONISATION
BETA RAYS β — IONISATION
GAMMA RAYS γ
X-RAYS X — IONISATION
NEUTRON RAYS n — IONISATION

PAPER | THIN ALUMINIUM | THICK LEAD | WATER OR CONCRETE

Beta particles
Because they're lighter and faster moving, beta particles have greater penetrating power. They pass through paper easily, but they can still be stopped by a sheet of aluminium just a few millimetres in thickness.

The electromagnetic spectrum
The EM spectrum spans a vast range of phenomena and technologies

Microwaves
Most commonly associated with microwave ovens, these frequencies – corresponding to wavelengths of a few centimetres – are also used by communication systems such as mobile phones and Wi-Fi. It's at this point that overexposure starts to become hazardous.

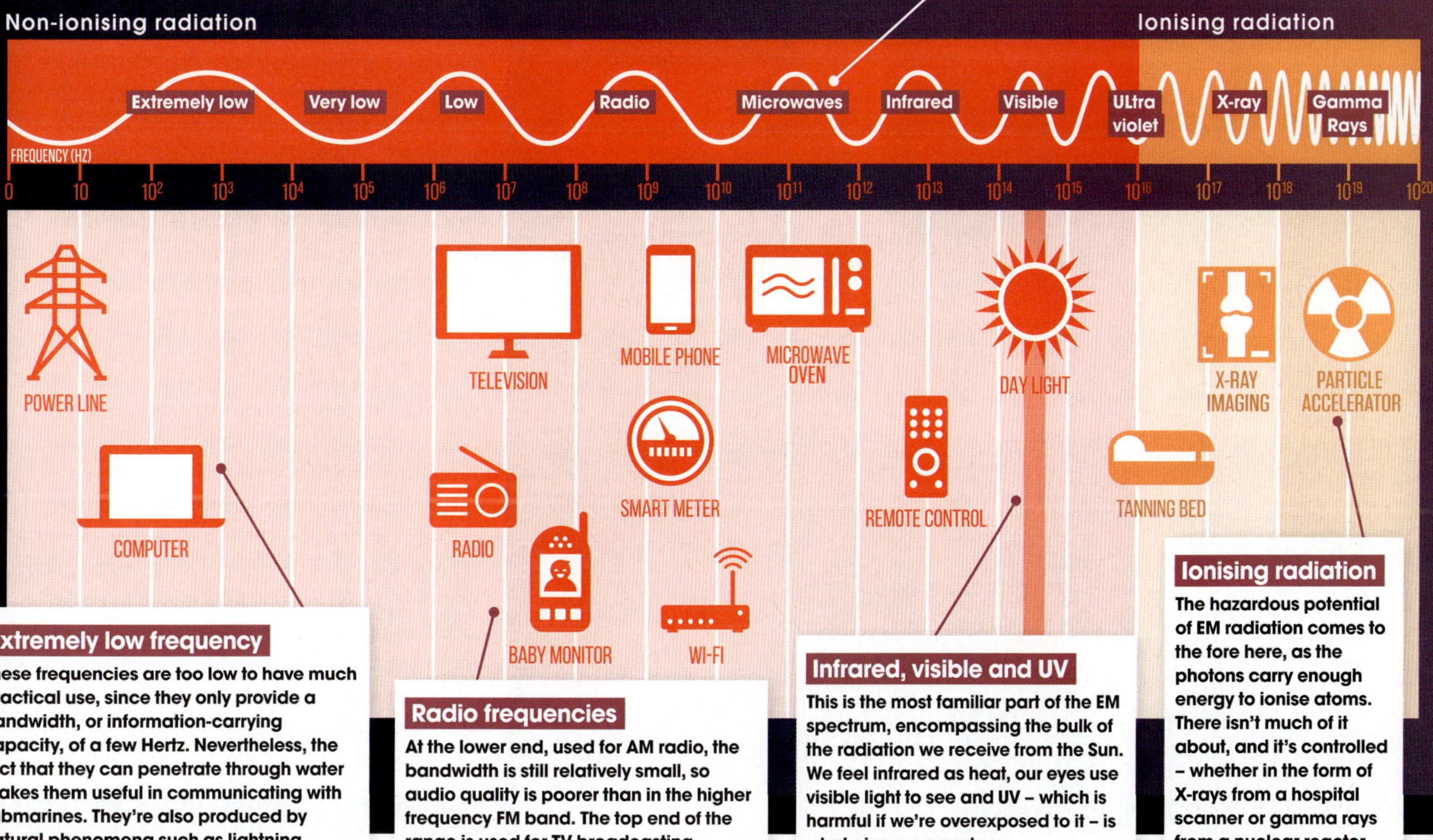

Non-ionising radiation | **Ionising radiation**

Extremely low | Very low | Low | Radio | Microwaves | Infrared | Visible | Ultraviolet | X-ray | Gamma Rays

FREQUENCY (HZ): 0, 10, 10^2, 10^3, 10^4, 10^5, 10^6, 10^7, 10^8, 10^9, 10^{10}, 10^{11}, 10^{12}, 10^{13}, 10^{14}, 10^{15}, 10^{16}, 10^{17}, 10^{18}, 10^{19}, 10^{20}

POWER LINE | COMPUTER | TELEVISION | MOBILE PHONE | MICROWAVE OVEN | RADIO | SMART METER | BABY MONITOR | WI-FI | DAY LIGHT | REMOTE CONTROL | X-RAY IMAGING | TANNING BED | PARTICLE ACCELERATOR

Extremely low frequency
These frequencies are too low to have much practical use, since they only provide a bandwidth, or information-carrying capacity, of a few Hertz. Nevertheless, the fact that they can penetrate through water makes them useful in communicating with submarines. They're also produced by natural phenomena such as lightning.

Radio frequencies
At the lower end, used for AM radio, the bandwidth is still relatively small, so audio quality is poorer than in the higher frequency FM band. The top end of the range is used for TV broadcasting.

Infrared, visible and UV
This is the most familiar part of the EM spectrum, encompassing the bulk of the radiation we receive from the Sun. We feel infrared as heat, our eyes use visible light to see and UV – which is harmful if we're overexposed to it – is what gives us a suntan.

Ionising radiation
The hazardous potential of EM radiation comes to the fore here, as the photons carry enough energy to ionise atoms. There isn't much of it about, and it's controlled – whether in the form of X-rays from a hospital scanner or gamma rays from a nuclear reactor.

095

UNDERSTANDING PHYSICS

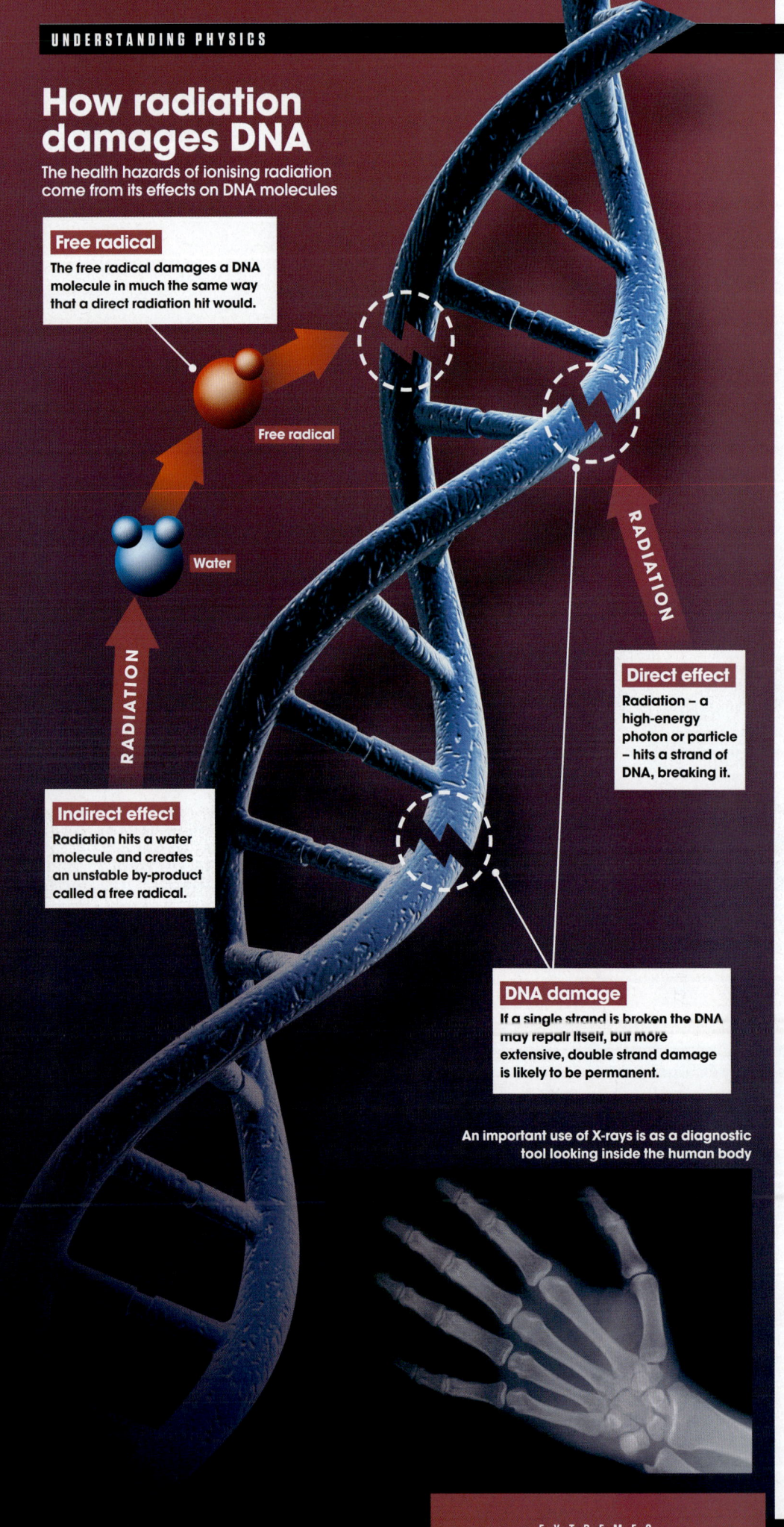

How radiation damages DNA

The health hazards of ionising radiation come from its effects on DNA molecules

Free radical
The free radical damages a DNA molecule in much the same way that a direct radiation hit would.

Free radical

Water

RADIATION

RADIATION

Direct effect
Radiation – a high-energy photon or particle – hits a strand of DNA, breaking it.

Indirect effect
Radiation hits a water molecule and creates an unstable by-product called a free radical.

DNA damage
If a single strand is broken the DNA may repair itself, but more extensive, double strand damage is likely to be permanent.

An important use of X-rays is as a diagnostic tool looking inside the human body

1895
Wilhelm Röntgen discovered ionising radiation in the form of X-rays

electrons to higher levels inside an atom, they knock them out of the atom altogether. Because electrons carry a negative electric charge, this turns the atom into a positively charged ion. This process is referred to as ionisation. Radiation capable of achieving this is called – predictably enough – ionising radiation, and it's what people are really thinking of when they associate radiation with danger.

X-rays, which come above UV in the EM frequency spectrum, are an example of ionising radiation. They can cause DNA damage and other medical problems – but only if the amount received is high enough. In small doses, such as you might receive in a dental X-ray or hospital scan, there's nothing to worry about. In fact, the best known property of X-rays – that they travel through soft tissue as if it was transparent – has been a huge boon to medical science.

When X-rays were first discovered by physicist Wilhelm Röntgen on 8 November 1895, he gave them that name – 'X' for unknown – because he had no idea what they were. On the other hand, it was obvious right away that they were incredibly useful. The first recorded use of X-rays for a medical diagnosis was on 11 January 1896, just nine weeks later. That's the quickest that a brand-new scientific discovery has ever found a practical application.

X-rays are produced by electronic means, just as microwaves, light and UV can be. But when we get to the top end of the EM spectrum, the photon energies are so huge they can only be created by processes inside the atomic nucleus. This is the fearsome gamma radiation that's produced, among other things, by nuclear bombs. But gamma rays don't have to originate in a tremendous explosion. At a much lower level, they're given off spontaneously by certain elements that have unstable nuclei in a process known as radioactivity. Some radioactive elements can only be produced artificially – for example in a nuclear explosion or a nuclear reactor – but others occur naturally. These natural sources of radioactivity, of which uranium is the best known, are around us all the time, albeit in relatively small quantities.

Classical scholars will be aware that gamma is the third letter of the Greek alphabet after alpha and beta. So where do alpha and beta radiation come into the picture? They're also forms of ionising radiation given off by radioactive substances, but they're not part of the EM spectrum. Rather than photons, they consist of

EXTREMES

DEADLY RADIATION

Everyday radiation
Not all doses are deadly – here's some of the radiation we experience

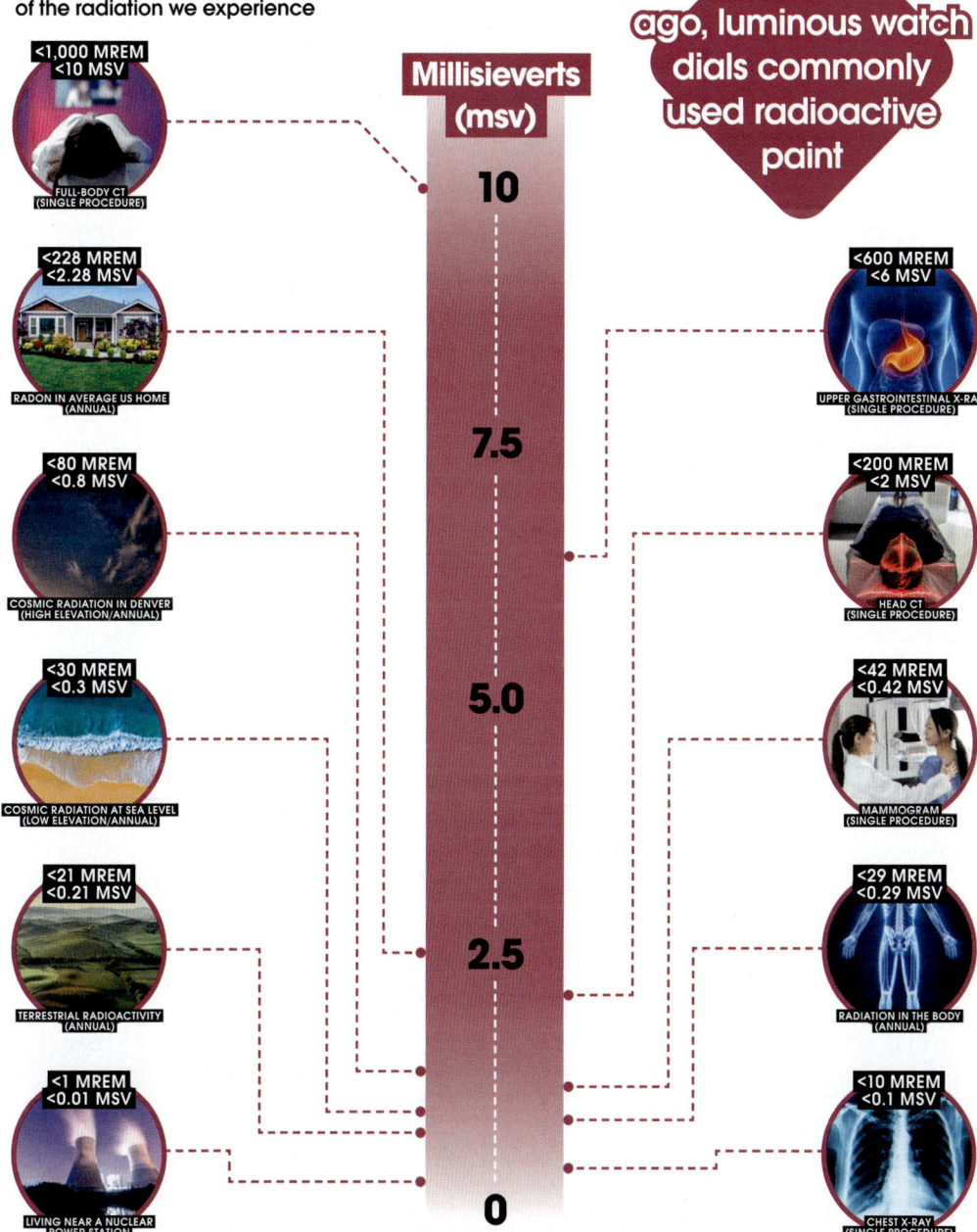

Millisieverts (msv)

- <1,000 MREM / <10 MSV — FULL-BODY CT (SINGLE PROCEDURE)
- <228 MREM / <2.28 MSV — RADON IN AVERAGE US HOME (ANNUAL)
- <80 MREM / <0.8 MSV — COSMIC RADIATION IN DENVER (HIGH ELEVATION/ANNUAL)
- <30 MREM / <0.3 MSV — COSMIC RADIATION AT SEA LEVEL (LOW ELEVATION/ANNUAL)
- <21 MREM / <0.21 MSV — TERRESTRIAL RADIOACTIVITY (ANNUAL)
- <1 MREM / <0.01 MSV — LIVING NEAR A NUCLEAR POWER STATION (ANNUAL)
- <600 MREM / <6 MSV — UPPER GASTROINTESTINAL X-RAY (SINGLE PROCEDURE)
- <200 MREM / <2 MSV — HEAD CT (SINGLE PROCEDURE)
- <42 MREM / <0.42 MSV — MAMMOGRAM (SINGLE PROCEDURE)
- <29 MREM / <0.29 MSV — RADIATION IN THE BODY (ANNUAL)
- <10 MREM / <0.1 MSV — CHEST X-RAY (SINGLE PROCEDURE)

A century ago, luminous watch dials commonly used radioactive paint

An early photograph, from 1915, of alpha particle tracks in a cloud chamber

Alpha and beta radiation

Unlike gamma rays, which are EM waves, the two other types of radiation emitted by radioactive substances consist of particles thrown out from unstable atomic nuclei. The nucleus is a dense clump of positively charged protons and electrically neutral neutrons. An alpha particle is composed of two neutrons and two protons, giving it a net positive charge and a relatively hefty mass. A beta particle is a fast-moving electron, which is much lighter and has a negative charge. Electrons are normally found in the outer parts of an atom, outside the nucleus, but beta particles are created by reactions inside the nucleus itself.

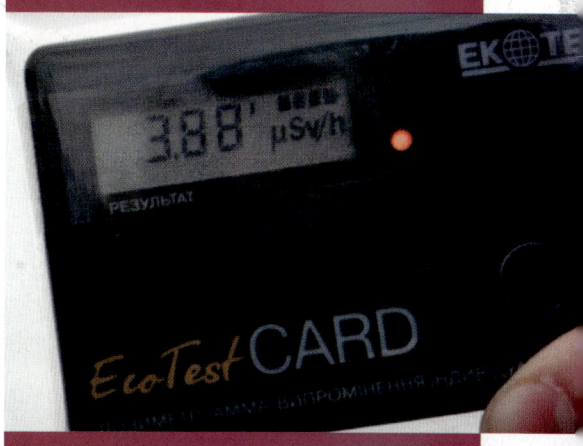

A dosimeter being used to check radiation levels at the site of the Chernobyl disaster

Measuring radiation

From a health-and-safety point of view, it's important to know how much ionising radiation is present in an environment. The most familiar way to measure this is the Geiger counter, which simply counts the number of ionising particles reaching it. But this isn't the best indication of danger level because it doesn't distinguish between high- and low-energy particles. A more refined alternative is the electronic dosimeter, which measures the cumulative energy received from ionising radiation in units called sieverts, with a fatal dose being about eight sieverts. For comparison, a dental X-ray gives you about five-millionths of a sievert, while you normally get twice that in the course of a day from natural background sources of radiation.

Even radio-frequency radiation can be hazardous right next to a powerful transmitter

WARNING
RADIO FREQUENCY RADIATION HAZARD
TO STOP RADIATION REMOVE KEY FROM PA INTERLOCK SYSTEM

097

UNDERSTANDING PHYSICS

streams of material particles: helium nuclei in the case of alpha rays and electrons in beta rays. What happens in radioactive decay is that an unstable nucleus spontaneously transforms into a more stable form, ejecting a photon or fast-moving particle in the process.

Alpha, beta and gamma rays were discovered in quick succession in the late 1800s and early 1900s, and were given their names by the 'father of nuclear physics', Ernest Rutherford. If you're interested in physics you've probably heard of him, but even if you aren't you'll be familiar with the name of one of his assistants at the University of Manchester, Hans Geiger. With Rutherford's help, he designed the first gadget for counting particles emitted by a radioactive sample. Geiger counters are still in use today, along with a range of more modern devices for measuring radiation levels – but the ominous and iconic clicks of a Geiger counter have become so familiar through movies and television that the term is often used loosely for all such devices.

If Geiger counters had existed centuries ago, they still would have given off the occasional click. That's because there's always a low background level of ionising radiation from natural sources. Some types of rock, such as granite, contain tiny traces of uranium and other radioactive elements. The Earth is also under constant bombardment from cosmic radiation, emanating both from the Sun and from more distant parts of the universe. This includes fast-moving protons and other high-energy particles, as well as X-rays and gamma rays. But the amount reaching the Earth's surface is much too small to pose any risk to life on Earth.

Since the second half of the 20th century, the level of background radiation has been boosted by the many nuclear weapons tests carried out between the 1940s and 1990s. The immediate radiation created by the explosions is long gone, but the blasts also produced a 'fallout' of radioactive material that has lingering effects to this day. The same is also true of the radioactive fallout produced by major incidents at nuclear power plants, such as Chernobyl in 1986 and Fukushima in 2011. These notoriously added to background radiation levels over a wide area.

Given the enormous potential of nuclear power to reduce worldwide carbon emissions, it's a tragedy when accidents like these cause so much damage to the environment. But the fact is that both Chernobyl and Fukushima were the result of human error – compounded, but not caused, in the latter case, by an earthquake and tsunami. If a nuclear power station is properly designed and correctly operated, there's no reason why it should leak any radiation into the environment at all.

1 million
Number of dental X-rays needed for a fatal radiation dose

Radiation in space

While EM radiation travels in straight lines, that's not always true of charged particles, which can be deflected by magnetism. This has a welcome effect on Earth, because the geomagnetic field shields us from the high-energy protons and electrons constantly streaming out from the Sun. But not all of that radiation is bounced harmlessly back into space – some of it gets trapped in doughnut-shaped rings around the Earth. Called the Van Allen belts, these start well above the altitude of the International Space Station, but they do pose a potential hazard for astronauts passing through them en route to more distant destinations.

Fortunately, a fast-moving spacecraft will only be inside the belts for an hour or so. In the case of the Apollo astronauts, NASA estimates they were exposed to 0.16 sieverts on their passage through the radiation belts – a relatively high dose, but still only about a fiftieth of the fatal level.

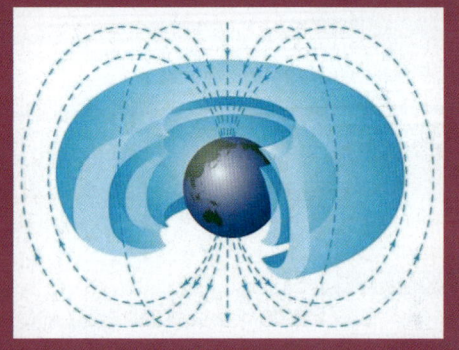

The Earth's magnetic field traps radiation in the doughnut-like Van Allen belts

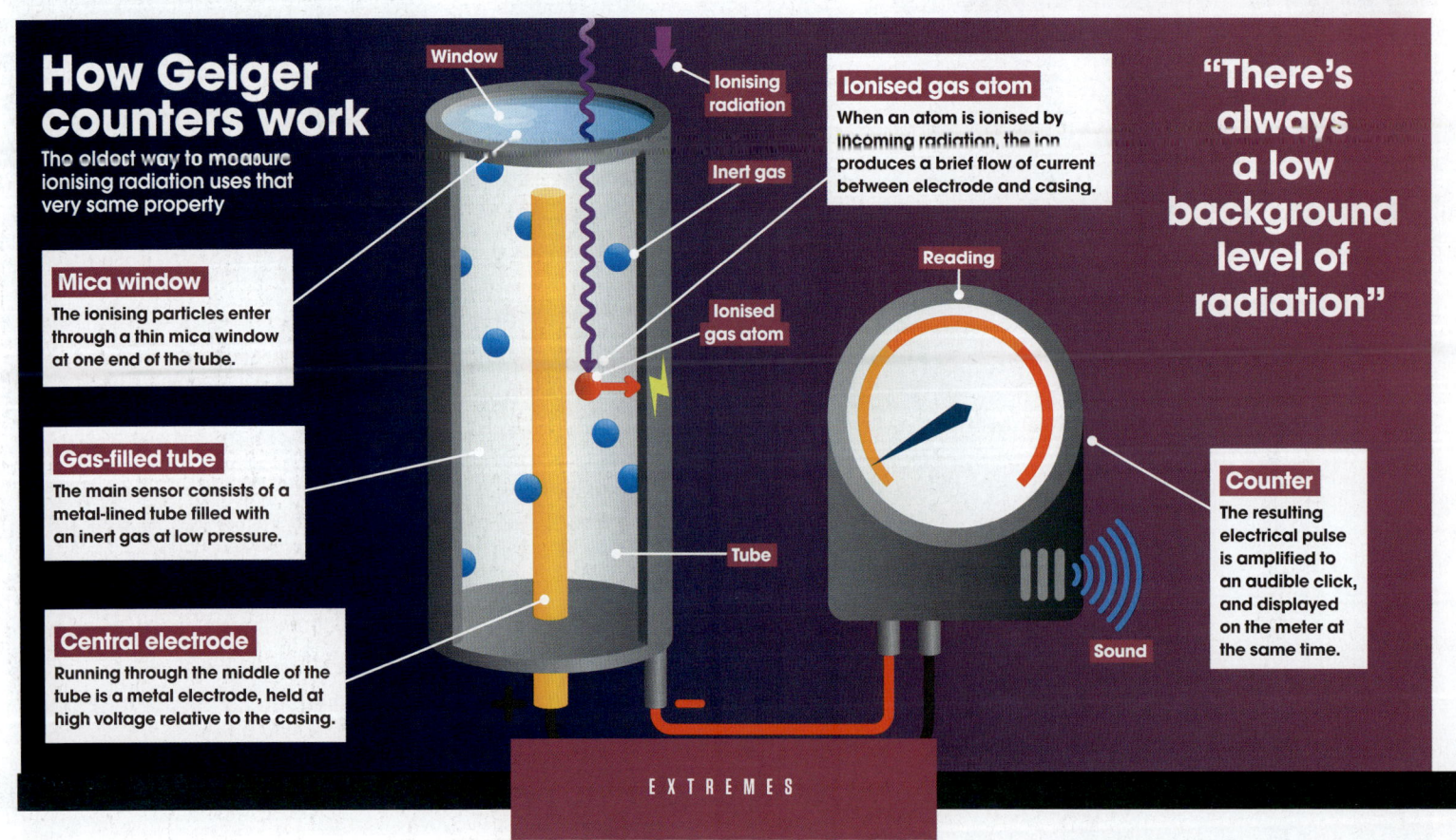

How Geiger counters work
The oldest way to measure ionising radiation uses that very same property

Mica window
The ionising particles enter through a thin mica window at one end of the tube.

Gas-filled tube
The main sensor consists of a metal-lined tube filled with an inert gas at low pressure.

Central electrode
Running through the middle of the tube is a metal electrode, held at high voltage relative to the casing.

Window

Ionising radiation

Ionised gas atom
When an atom is ionised by incoming radiation, the ion produces a brief flow of current between electrode and casing.

Inert gas

Ionised gas atom

Reading

Tube

Counter
The resulting electrical pulse is amplified to an audible click, and displayed on the meter at the same time.

Sound

"There's always a low background level of radiation"

EXTREMES

DEADLY RADIATION

Water is excellent at blocking radiation due to its hydrogen molecules. This is the Advanced Test Reactor in Idaho. It's submerged in water, generating the blue glow of Cherenkov radiation

UNDERSTANDING PHYSICS

The Hidden Universe

Dark matter and dark energy make up 95 per cent of the universe, yet we can't see them. What is this strange stuff?

EXTREMES

THE HIDDEN UNIVERSE

As telescopes became increasingly powerful during the 20th century, they started to reveal the true scale of the cosmos. Astronomers discovered that there were billions of other galaxies like our own, scattered throughout a vast, continuously expanding universe. At the same time, advances were made in theoretical cosmology, stemming from Einstein's theory of general relativity, which showed in precise detail how objects move under the influence of gravity. When those two developments – observational and theoretical – were put together, researchers came to a startling conclusion. By the end of the 20th century, it was clear that all those billions of visible galaxies were just a small fraction of everything there is.

The hidden 95 per cent of the universe goes by the names dark matter and dark energy – but these are two very different things. The word dark is appropriate in the sense that we are 'in the dark' about them – we can't observe them directly, and we don't know what they are. But it's misleading to think of them as being dark in colour. That's true of something like cosmic dust, which we can see quite easily if it gets between us and a bright object that it partially obscures, but dark matter and dark energy are completely transparent. Light across all wavelengths, and all other matter, simply passes through them as if they weren't there.

Dark matter was discovered first – and the underlying theory is easier to understand. There's no need for relativity here, just Isaac Newton's basic theory of gravity. When you have a large ensemble of stars in a galaxy – or galaxies in a galaxy cluster – despite all the complex physics that's going on inside them, gravity is the only thing that determines their motion. Just as a spacecraft can attain 'escape velocity' from Earth orbit if it's moving fast enough, there's a maximum speed that stars can travel at – determined by the total amount of gravitating matter in the galaxy – before they fly off at a tangent. It turns out that the stars in the outer parts of most galaxies are moving too fast, at least if the visible matter was the only thing holding them in. The concept of dark matter, which supplies the missing gravity but is undetectable by any other means, is the simplest way to explain the observations.

Astronomers see evidence for dark matter everywhere they look – here in our own galaxy, and in other neighbouring galaxies. In contrast, dark energy only becomes apparent when we

A dense galaxy cluster interspersed with blue arcs, which are more distant, gravitationally lensed galaxies

The constituents of the universe

Relative proportions of dark energy and matter – the latter split further into 'ordinary' and 'dark'

From observations of stellar and galactic motions, astronomers know the universe must contain around five times as much dark matter as ordinary visible matter. Adding dark energy to the picture is a little harder. It isn't made up of material particles, as dark matter presumably is, so we can't simply characterise its contribution as so many kilograms per cubic metre. But thanks to Einstein's theory of relativity, we know that energy is equivalent to mass, and cosmological observations allow us to work out the amount of dark energy in a way that is directly comparable to the other two. The result, according to NASA's latest estimate, is that the universe is 68 per cent dark energy, 27 per cent dark matter and just five per cent ordinary matter.

UNDERSTANDING PHYSICS

take a wider view of the universe as a whole. For a century now we've known that the universe has been expanding ever since the Big Bang. It's common sense to assume that this expansion is gradually getting slower over time, pulled back by the combined gravity of all the matter in the universe. But in the 1990s astronomers discovered that the exact opposite is true: the expansion rate is actually accelerating, not slowing down. Something is counteracting the effect of gravity, pushing galaxies apart faster and faster. That 'something' – and no one can prove exactly what it is yet – has been dubbed "dark energy".

One thing both dark matter and dark energy have in common is an absence of direct evidence. They're assumed to exist because they're the simplest way to reconcile observations with theory. But it's possible that theory and observations are wrong, and we don't really need dark matter or dark energy after all. But the indirect evidence for them is mounting up all the time, so most astronomers believe they're here to stay.

Dark matter

It was in the 1930s that Fritz Zwicky first noticed a discrepancy between the visual appearance of galaxies and the speeds they were travelling at. When studying the Coma galaxy cluster, he realised that in order for it to be held together by gravity, it had to contain far more mass than he could see. He coined the term 'dark matter' for the unseen contribution.

By the 1960s, spectroscopy had progressed to the point where high-resolution measurements could be made of stellar velocities inside a galaxy and plotted against radius. One of the great pioneers of these 'galactic rotation curves' was Vera Rubin. She discovered that the outer parts of most disc galaxies rotate much faster than would be expected from the gravitational effect of visible matter. The implication was that galaxies were embedded in a 'halo' of dark matter, the density of which dropped off more slowly with radius than that of the visible disc.

Disc galaxies, like the Sculptor Galaxy shown here, are embedded in a halo of dark matter

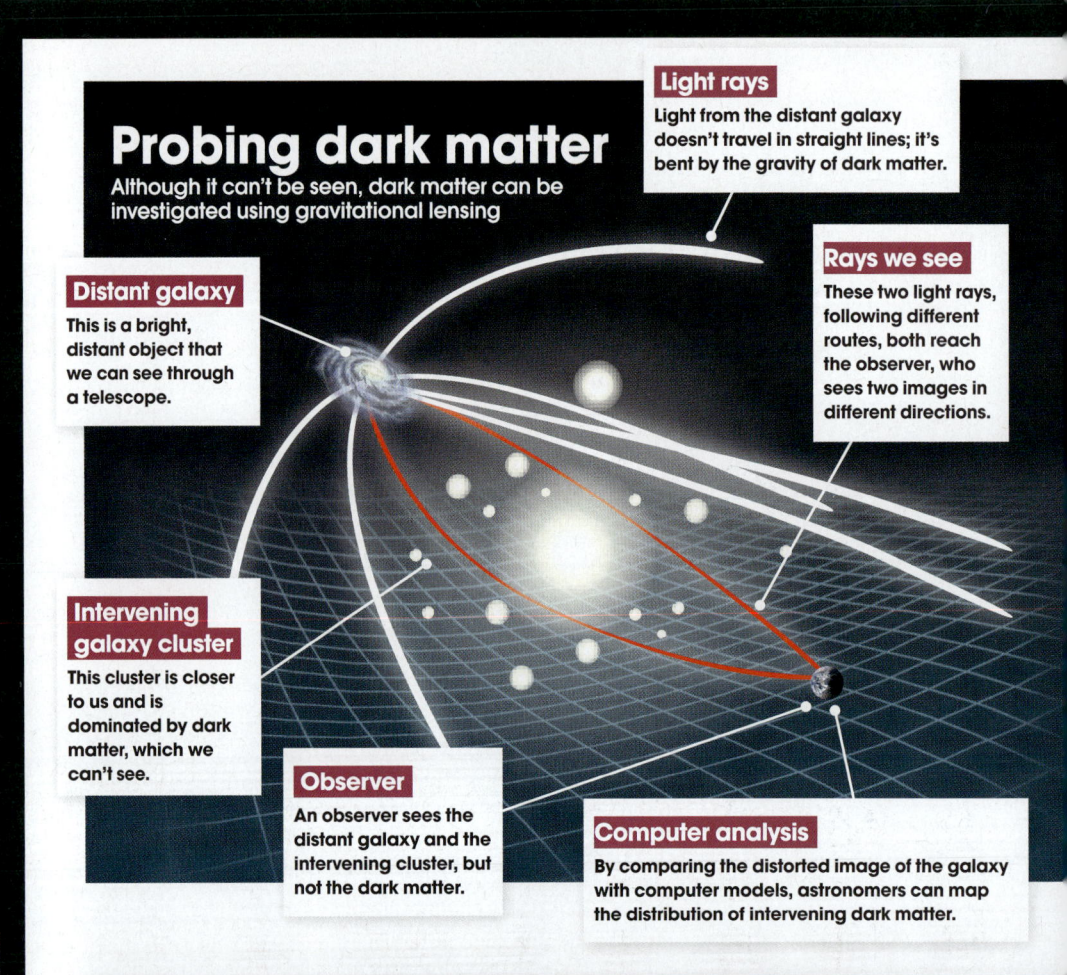

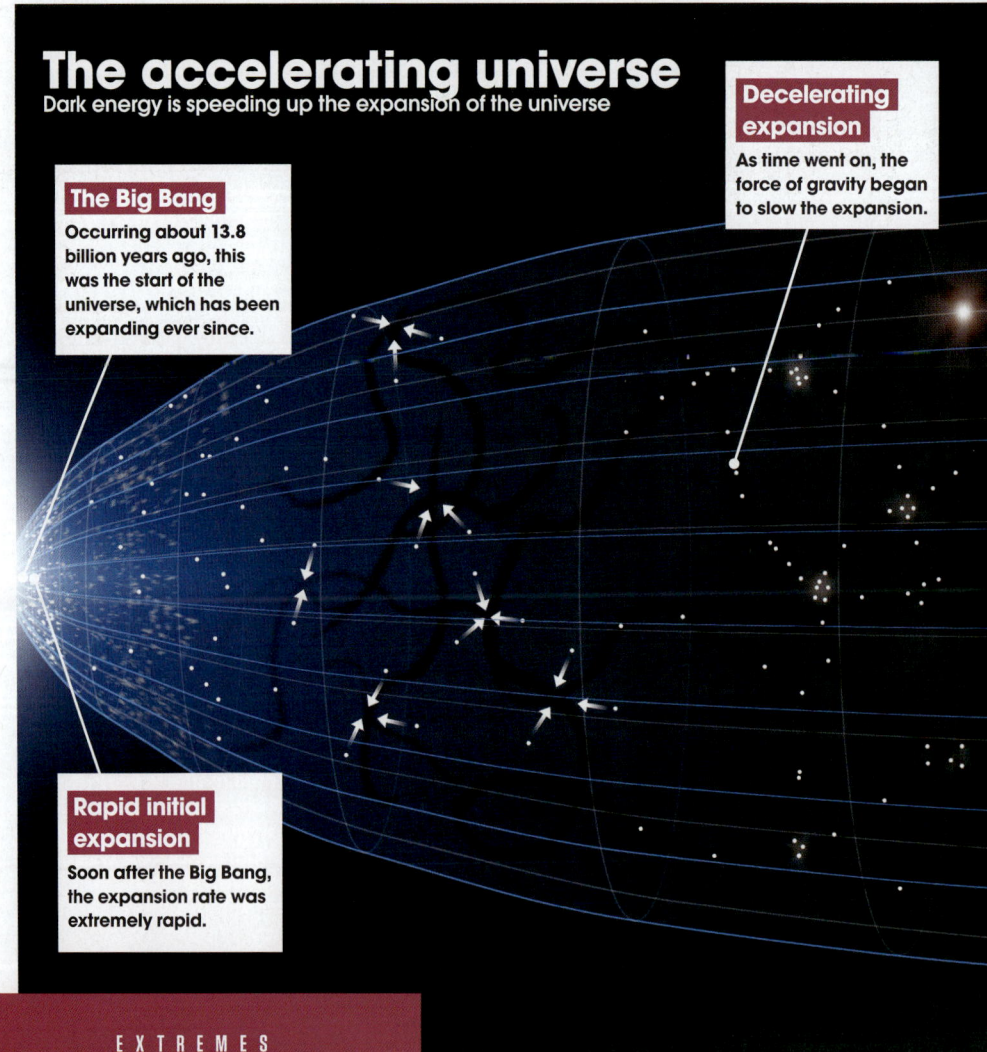

Probing dark matter
Although it can't be seen, dark matter can be investigated using gravitational lensing

Light rays
Light from the distant galaxy doesn't travel in straight lines; it's bent by the gravity of dark matter.

Rays we see
These two light rays, following different routes, both reach the observer, who sees two images in different directions.

Distant galaxy
This is a bright, distant object that we can see through a telescope.

Intervening galaxy cluster
This cluster is closer to us and is dominated by dark matter, which we can't see.

Observer
An observer sees the distant galaxy and the intervening cluster, but not the dark matter.

Computer analysis
By comparing the distorted image of the galaxy with computer models, astronomers can map the distribution of intervening dark matter.

The accelerating universe
Dark energy is speeding up the expansion of the universe

Decelerating expansion
As time went on, the force of gravity began to slow the expansion.

The Big Bang
Occurring about 13.8 billion years ago, this was the start of the universe, which has been expanding ever since.

Rapid initial expansion
Soon after the Big Bang, the expansion rate was extremely rapid.

EXTREMES

THE HIDDEN UNIVERSE

The Euclid mission

The European Space Agency's (ESA) Euclid space telescope, scheduled for launch next year, is designed to investigate both dark matter and dark energy. It will map gravitationally lensed galaxies, from which the distribution of intervening dark matter can be deduced. It will also study so-called 'baryonic acoustic oscillations', which are ancient patterns imprinted in the large-scale distribution of galaxies. Like explosive supernovae, these provide a standard ruler which allows astronomers to trace the expansion history of the universe – including the acceleration caused by dark energy. In a mission lasting six years, Euclid will survey galaxies in visible and infrared wavebands over an area of sky covering more than 35 per cent of the celestial sphere.

An artist's impression of the Euclid spacecraft in its operational configuration

Dark energy

Astronomers accidentally discovered dark energy when they were looking for something else. They wanted to calculate the total mass of the universe by measuring the rate at which its gravitational pull slowed down the expansion rate. They attempted to do this by graphing recession speed versus distance for a special class of astronomical objects called Type Ia supernovae, but the result wasn't what they expected. The expansion of the universe isn't slowing down at all – in fact, it's speeding up.

It's as though the universe is filled with a mysterious something – dark energy – that counteracts the pull of gravity even on the largest of scales and pushes even harder in the opposite direction. This discovery was in a different league from dark matter, which may be a completely unknown substance, but at least it obeys the established laws of Newton and Einstein. Dark energy, with its strange, antigravity-like behaviour, doesn't even do that.

Or perhaps it does, if we look at an obscure factor in Einstein's equation called the cosmological constant. It has no counterpart in Newton's theory, and for years it was assumed to be zero. But if it has a small positive value, it could explain the existence of dark energy as a fundamental property of space itself.

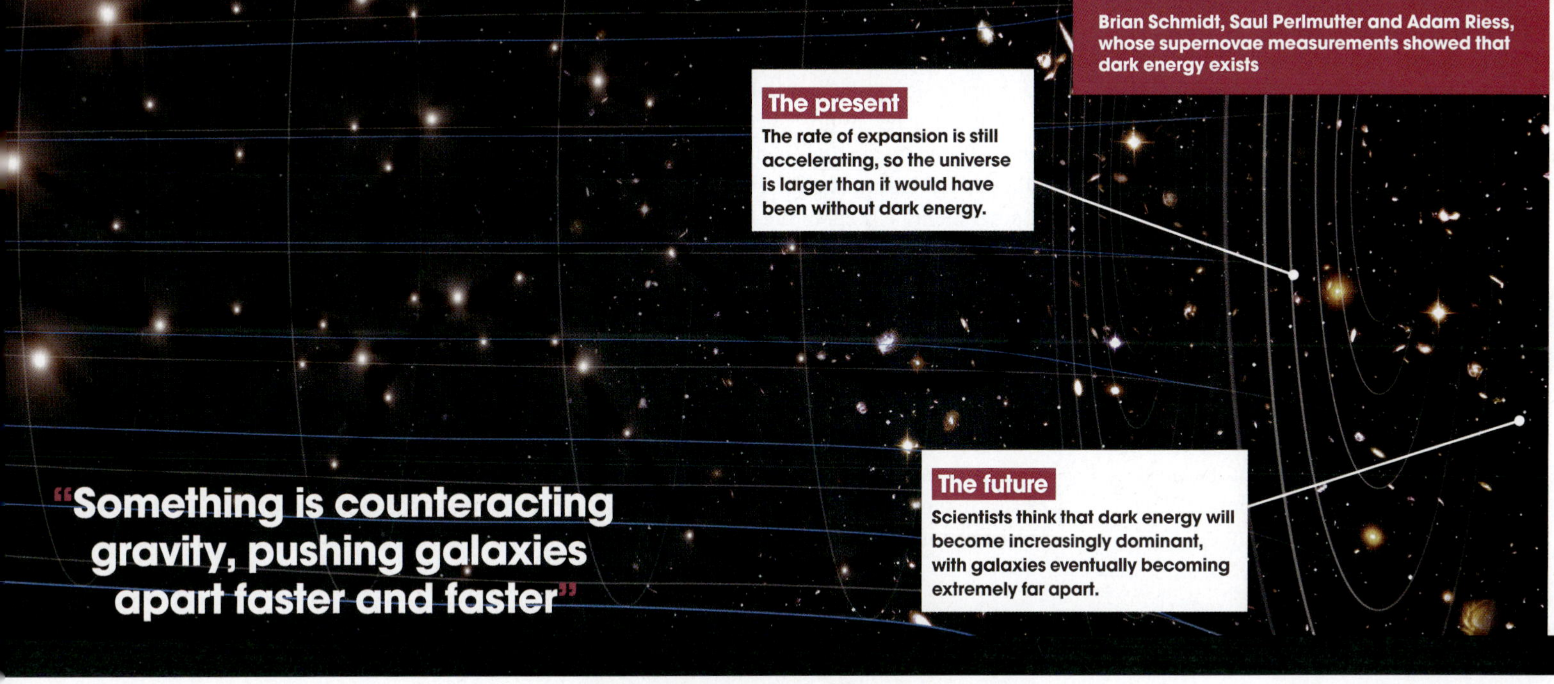

Dark energy kicks in
Around 5 billion years ago, dark energy started to affect expansion, which began to speed up again.

Einstein included something similar to dark energy – the cosmological constant – in his theory of relativity

Brian Schmidt, Saul Perlmutter and Adam Riess, whose supernovae measurements showed that dark energy exists

The present
The rate of expansion is still accelerating, so the universe is larger than it would have been without dark energy.

The future
Scientists think that dark energy will become increasingly dominant, with galaxies eventually becoming extremely far apart.

"Something is counteracting gravity, pushing galaxies apart faster and faster"

UNDERSTANDING PHYSICS

The power of atoms

Atoms are the ultimate construction kit. With a big enough collection, you can build everything from Venus de Milo to Venus the planet...

Everything in the universe is made of countless tiny atoms that are joined into different structures, essentially like toy building blocks. But instead of snapping together through friction, like plastic blocks, atoms snap together through electrical charge.

At the centre of every atom you'll find the nucleus – a blob of positively charged particles called protons and neutrally charged particles called neutrons. The tiny nucleus is surrounded by even smaller negatively charged particles called electrons. Usually, there is an equal number of protons and electrons. Because of their opposite charge, these protons and electrons are attracted to each other, which is what holds the atom together. With this balance, the atom is electrically neutral.

But these electrons are a fickle bunch: they're not only attracted to their own atom's nucleus – they're sometimes attracted to the nuclei of other atoms as well. In the right situations and conditions, this cross-atom attraction can provide a sort of 'electron glue' that bonds multiple atoms together.

An atom's bonding prospects depend on its proton and electron count and arrangement, which is unique for every element on the periodic table. Electrons surround the nucleus at specific energy levels, called shells. The shell closest to the nucleus is the lowest energy level, and the shell farthest from the nucleus is the highest energy level. Each shell can hold a limited number of electrons. For example, the lowest-level shell holds a maximum of two electrons, and the next level holds up to eight electrons. To achieve maximum stability, electrons move to the lowest possible energy level that has available openings.

The critical factor in chemical bonding is the number of openings in an atom's outermost shell, called the valence shell. When there is the right combination of openings, electrons can jump from one atom to another, two atoms can share an electron, or many atoms can share a cloud of electrons. Atoms are more stable when their valence shells are full, so electrons will readily move in ways that form complete

EXTREMES

THE POWER OF ATOMS

At full power, trillions of protons will race around the LHC accelerator ring 11,245 times a second

Anatomy of an atom
What are the fundamental parts that make an atom?

Shells
Electrons can only exist in set energy levels, commonly called shells. Each shell has openings for a limited number of electrons.

Electrons
Electrons are very small, negatively charged particles that move quickly around the atom's nucleus.

The nucleus
The centre of the atom, and almost all of its mass, is the nucleus. The nucleus is made up of protons and neutrons.

Protons
Protons are positively charged particles in the nucleus. All elements are defined by how many protons they have.

Neutrons
Neutrons – particles with no electrical charge – help give atoms their mass. They are slightly bigger than protons.

The 'Fat Man' nuclear bomb was dropped on Nagasaki, Japan, on 9 August 1945

---

valence shells.

When multiple atoms bond together, they form molecules. Molecules can consist of many identical atoms – that is, atoms of the same element – or they can include atoms of multiple elements. A multi-element molecule is called a compound. Collectively, vast numbers of molecules form the wide variety of materials that we know and love today. The structure of these individual molecules, along with the way that those molecules fit together, ultimately dictates how any material both feels and behaves.

Broadly speaking, there are three styles of organisation: gases, liquids and solids. In gases, molecules move about freely. In liquids, molecules fit together loosely, sliding over one another like marbles in a bowl. In solids, meanwhile, molecules are arranged in more rigid structures, and so don't move as freely.

Within these groups, different combinations and arrangements of atoms result in an incredible range of qualities and behaviours. Even limited to a set of identical atoms, structural changes can make huge differences. For example, compare diamonds and graphite. Both are arrangements of carbon atoms, but you don't see anyone proposing on bended knee with a pencil.

In a diamond, strong covalent bonds join atoms in a rigid lattice framework. The result is one of the hardest, toughest materials in the world. In graphite, on the other hand, carbon atoms are arranged in a layered structure, with very weak bonds between layers – so weak that touching a pencil to paper is enough to break them. The ability to combine and arrange atoms into different structures yields virtually unlimited possibilities. Scientists and engineers have already developed countless thousands of novel materials, and we're nowhere close to exhausting the potential combinations.

The chemical reactions involved in recombining atoms can prove useful themselves. For example, fire is the result of a chemical reaction between the chemical compounds in wood (or some other fuel) and oxygen in the atmosphere, triggered by intense heat. Burning wood produces char and gaseous compounds of hydrogen, carbon and oxygen. As the gases heat up, the compounds break apart, and the atoms recombine with oxygen in the air to produce water, carbon dioxide, carbon monoxide and nitrogen; this releases a great deal of energy in the form of heat and light in the process.

Between forming new materials and producing usable energy, the manipulation of atoms has always been at the heart of human technology – even when we had no clue atoms existed. In recent years, scientists have managed to make new atoms, forming 20 elements not observed in nature by combining existing nuclei into new super-heavy nuclei. These manmade atoms quickly fall apart, but stable variations may not be far off. In the 20th century, humans unlocked the internal energy of atomic nuclei for the first time, yielding both nuclear powerplants and bombs.

Today, physicists are investigating the even smaller components – quarks, leptons and bosons – that make up atoms. At this still mysterious level, new findings could fundamentally redefine our understanding of the universe.

UNDERSTANDING PHYSICS

Atomic models

Atoms don't follow the rules of Newtonian physics that we see every day in the world around us, making it impossible to visualise what's actually happening at an atomic level. The best that scientists can do is create theoretical models that give us a general conceptual comprehension of what's going on. Here are some of the noteworthy models that greatly advanced our atomic understanding...

Thomson's 'Plum Pudding' model (1904)

English physicist JJ Thomson discovered the electron as far back as 1897, showing for the first time that atoms had smaller constituent components. To account for the atom's overall neutral charge, Thomson theorised – in his 1904 model – that the negatively charged electrons must sit in a regular pattern within a uniformly distributed positive charge, like raisins in a plum pudding.

Rutherford's nuclear model (1911)

New Zealand-born physicist Ernest Rutherford, who studied at Cambridge University, disproved Thomson's model, when he demonstrated the existence of a positively charged atomic nucleus. Rutherford proposed the atom was like a miniature solar system, with a relatively massive, Sun-like nucleus at the centre, orbited by much smaller planet-like electrons.

Bohr's shell model (1913)

In classical mechanics, any charged particle moving in a curved path emits radiation. Consequently, in Rutherford's model, electrons would lose energy and collapse into the nucleus. Danish physicist Niels Bohr proposed electrons moved in a different type of orbit. He theorised electrons surrounded a nucleus in fixed energy levels (shells) and only emitted radiation when they 'jumped' from shell to shell.

What are elements and compounds?

Elements are substances made entirely of one type of atom. Each element is defined by how many protons are in a single atom of that element. For example, every hydrogen atom has just one proton, while every gold atom has 79 protons, and so on. In the right circumstances, atoms of different elements can join together to form a chemical compound. The bonds that hold compounds together result from various movements of electrons. Here are two examples:

Ionic bond

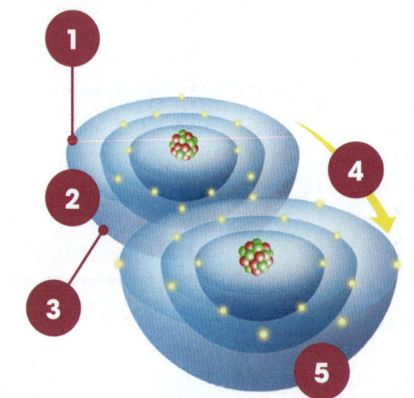

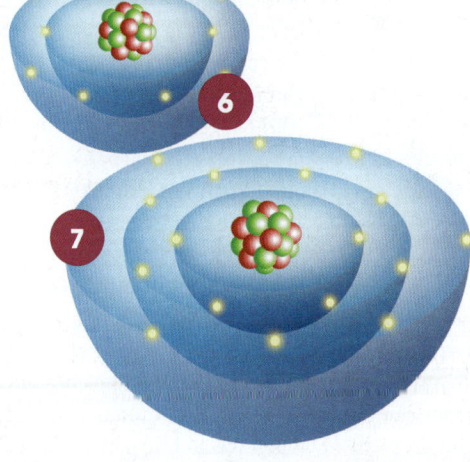

1 Ionic bond
Ionic bonds form when an electron jumps from one atom to another, resulting in two electrically charged atoms, called ions.

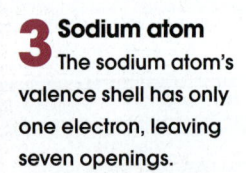

3 Sodium atom
The sodium atom's valence shell has only one electron, leaving seven openings.

5 Chlorine atom
A chlorine atom's valence shell holds seven electrons, leaving only one opening.

6 Sodium cation
The sodium atom now has ten electrons and 11 protons, making it a cation – an atom with a net positive charge.

Covalent bond

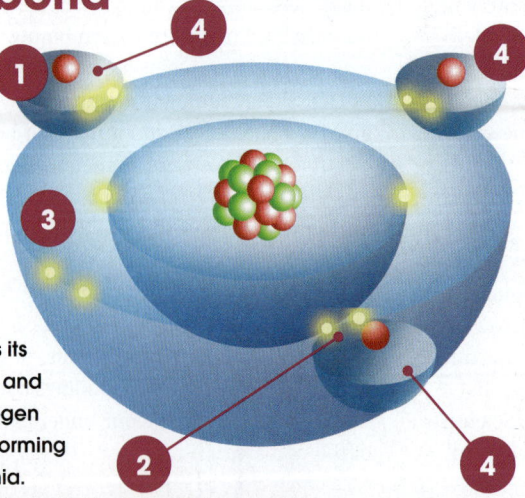

1 Covalent bond
Atoms can also form compounds by sharing electrons between them, in covalent bonds.

2 Electron pairs
Each of the three hydrogen atoms shares its original single electron and one of the original nitrogen electrons, collectively forming the compound ammonia.

2 Valence shell
Electrons travel in set energy levels called shells. Each shell has a limited number of openings for electrons. The number of openings in the outermost level, known as the valence shell, determines how an atom can form bonds.

4 Electron leap
To achieve overall stability, the spare electron from the sodium atom leaps to fill the chlorine atom's valence shell.

7 Chlorine anion
The chlorine atom now has 18 electrons and 17 protons, making it an anion – an atom with a net negative charge. The opposite charges bond the two atoms together to form the compound sodium chloride, more commonly known as table salt.

3 Nitrogen atom
A nitrogen atom has five electrons in its valence shell, leaving three openings.

4 Hydrogen atoms
Each of these three hydrogen atoms has a single electron in its valence shell, leaving one opening.

EXTREMES

THE POWER OF ATOMS

Bombarding atoms together leads to the dislodging of subatomic particles

From stars to space dust, everything in the universe is made up of atoms

How to split an atom

Add this to the pile of mind-bending atom qualities: an atomic nucleus has less mass as a whole than its protons and neutrons would have separately. How is this possible? Well, when the nucleus is formed, some of the mass of its constituent parts changes into energy that binds the protons and neutrons together. In other words, there's high potential energy locked up in the nucleus of the atom.

It's possible to release this energy, and actually harness it, by splitting specific types of atoms apart into multiple fragments – a process known as nuclear fission. All you need to break apart a uranium-235 atom is a slow-moving free neutron. The uranium atom will absorb the free neutron, the extra energy makes the uranium nucleus highly unstable, and the atom splits into two smaller atoms and two or three free neutrons. The potential energy in the nucleus is released as kinetic energy, in the form of these particles moving at great speed. The resulting free neutrons, in turn, can break apart other uranium-235 atoms, leading to a chain reaction.

A powerplant controls the reaction and harnesses the heat of this kinetic energy in order to generate steam that turns turbines. In contrast, in an atomic bomb the reaction is allowed to go unchecked, in order to generate a massive explosion.

You can also tap into this energy through nuclear fusion – the combining of two nuclei into a new nucleus. Nuclear fusion generates the energy of stars and hydrogen bombs. However, nobody has been able to harness it effectively as a power source yet.

> "All you need to break apart a uranium-235 atom is a slow-moving free neutron"

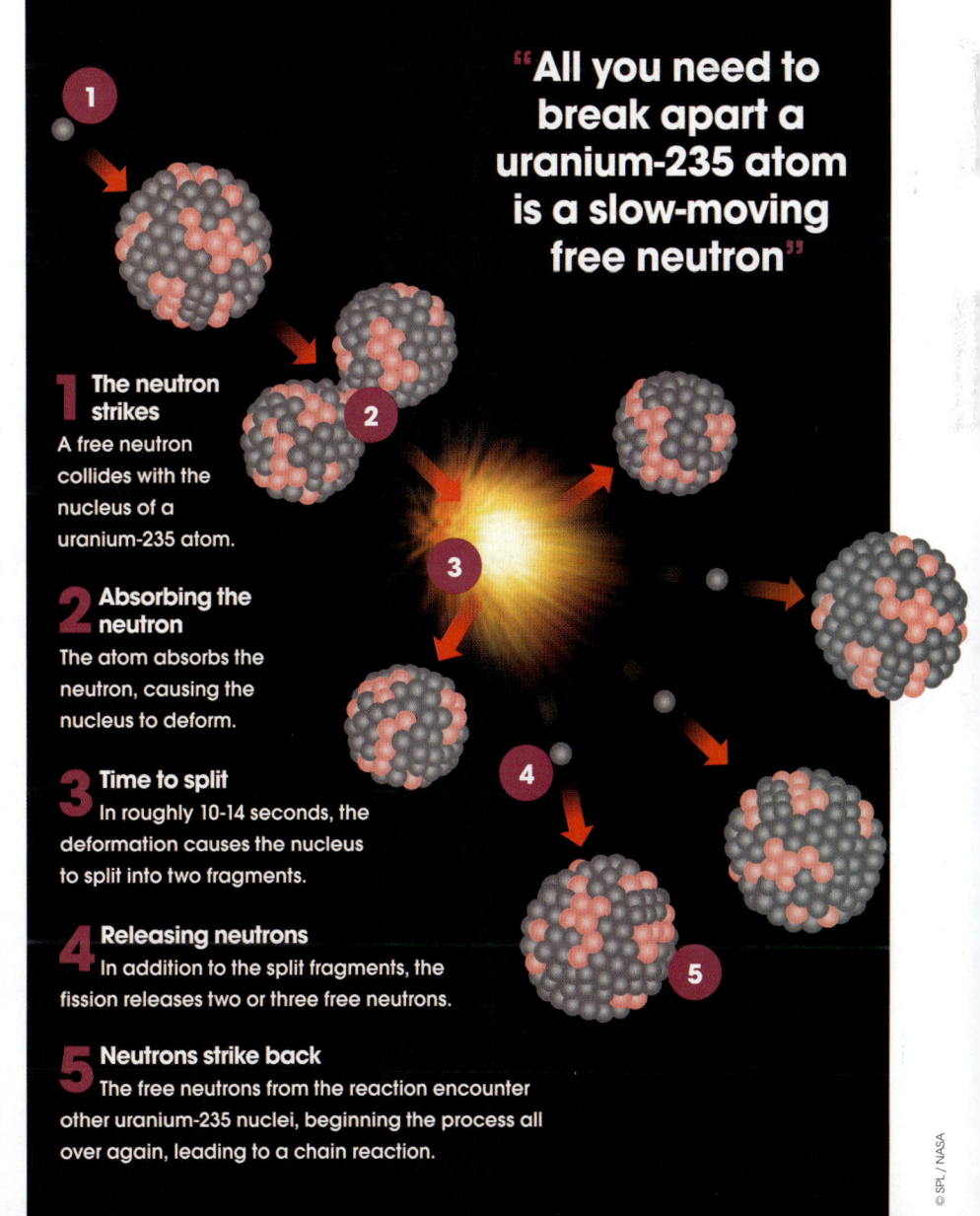

1 The neutron strikes
A free neutron collides with the nucleus of a uranium-235 atom.

2 Absorbing the neutron
The atom absorbs the neutron, causing the nucleus to deform.

3 Time to split
In roughly 10-14 seconds, the deformation causes the nucleus to split into two fragments.

4 Releasing neutrons
In addition to the split fragments, the fission releases two or three free neutrons.

5 Neutrons strike back
The free neutrons from the reaction encounter other uranium-235 nuclei, beginning the process all over again, leading to a chain reaction.

Radioactive decay explained

Most atoms are highly stable, meaning that the nucleus will always hold together, barring extreme circumstances. But in some atoms, the energy that binds the nucleus will eventually fail in a process called radioactive decay – the spontaneous disintegration of the nucleus.

The most notorious unstable atoms are elements with a very high number of protons, such as uranium (92 protons). But some lighter elements, such as carbon, can be unstable and radioactive as well, in instances when they have too many or too few neutrons. Neutron-count variations are called isotopes. For example, while garden-variety carbon-12 (six protons and six neutrons) is entirely stable, carbon-14 (six protons and eight neutrons) is radioactive.

Radioactive decay results in the ejection of subatomic particles from the nucleus. In alpha radiation, the atom ejects two protons and two neutrons. In beta radiation, a neutron turns into a proton, ejecting a neutrino particle and a free electron, called a beta wave. In gamma radiation, the nucleus releases extra energy in the form of a photon. Energy from the ejected particles can damage and mutate human DNA, sometimes resulting in disastrous cellular changes, ie cancer.

The nucleus material that's left forms a 'daughter atom'. When the proton count has changed, the daughter atom will be entirely different from the original atom. Carbon-14 decays into nitrogen, for example.

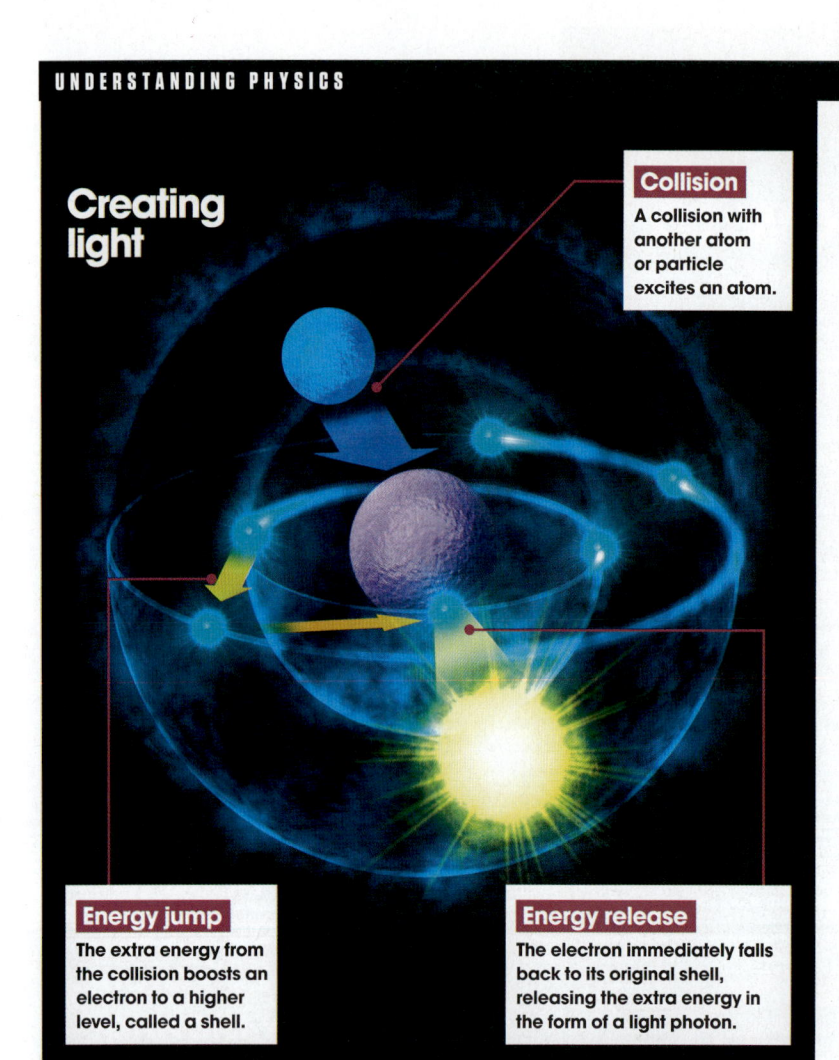

UNDERSTANDING PHYSICS

Creating light

Collision
A collision with another atom or particle excites an atom.

Energy jump
The extra energy from the collision boosts an electron to a higher level, called a shell.

Energy release
The electron immediately falls back to its original shell, releasing the extra energy in the form of a light photon.

How atoms emit light

Light is the result of electrons moving between defined energy levels in an atom, called shells. When something excites an atom, such as a collision with another atom or a chemical electron, an electron may absorb the energy, boosting it up to a higher-level shell. The boost is short-lived, however, and the electron immediately falls back down to the lower level, emitting its extra energy in the form of an electromagnetic energy packet called a photon. The wavelength of the photon depends on the distance of the electron's fall. Some wavelengths, such as radio waves, are invisible. Photons with wavelengths in the visible spectrum form all the colours that we can see.

The D-Wave One from D-Wave Systems is the first commercially available quantum computing system

Atoms and quantum computing

One of the many oddities about activity on the subatomic level is that subatomic particles do not have a defined state until they are observed. Instead of saying exactly where a proton, electron or other subatomic particle actually is, physicists instead talk about something called a probability cloud, indicating all of its possible states.

The weird but real phenomenon of quantum tunnelling helps illustrate this. As a subatomic particle approaches a barrier, one edge of the probability cloud for its position moves to the other side of the barrier. So, there's a small chance it actually will be on the other side of the barrier. Sometimes, it is on the other side, effectively tunnelling straight through the barrier.

Another way to define this ambiguity is to say a subatomic particle is in all possible positions at once. Contrast this with a computer bit, which at any moment, has a value of either 1 or 0. The fundamental idea of quantum computing is to employ each of the many 'superposed' states to perform part of a calculation, in order to do the entire calculation far more quickly than a conventional computer could manage. The field is now in its infancy, with limited implementations, but it could revolutionise computing in the foreseeable future.

> "13.5 per cent of the world's electricity in 2010 came from the world's 436 nuclear reactors"

France is currently leading the way with nuclear power, with nearly 80 per cent of its overall energy derived from atom power

Atom power by the numbers

No matter how you feel about nuclear power, it's part of your life. According to the Nuclear Energy Institute, 13.5 per cent of the world's electricity production in 2010 came from the world's 436 nuclear reactors. France leads the pack, drawing 77.7 per cent of its power from nuclear energy in 2011. The UK comes in at considerably less at 15.7 per cent.

Other energy sources still exceed nuclear power. The International Energy Agency lists coal/peat as the top source, providing 40.6 per cent of the world's power. Natural gas is second with 21.4 per cent, followed by hydropower with 16.2 per cent. Solar, wind, biofuels, heat and geothermal power combined amount to only 3.3 per cent.

EXTREMES

THE POWER OF ATOMS

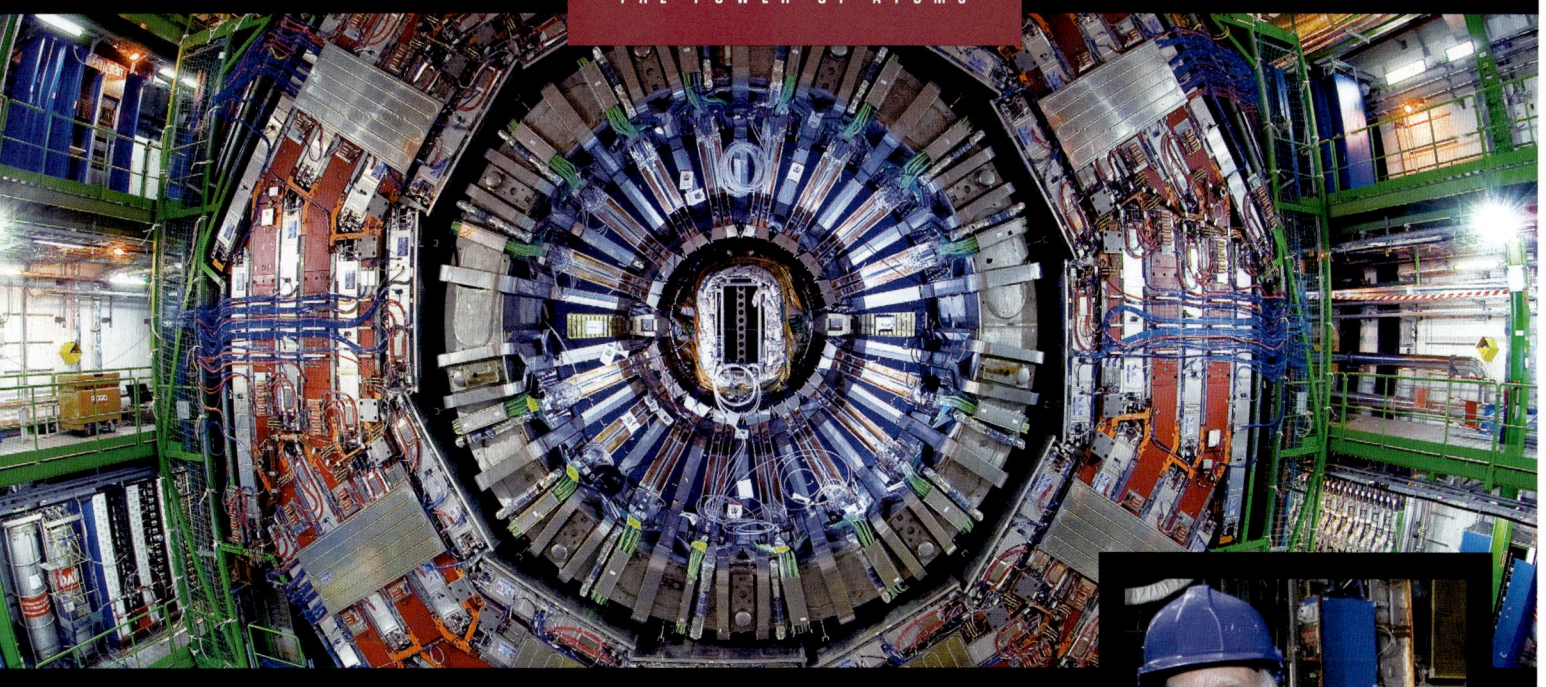

What is the Higgs boson?

The Higgs boson is a theoretical particle proposed by British physicist Peter Higgs as part of the Standard Model of particles and forces. The Standard Model is an incomplete theory that describes how the 12 known fundamental particles and three of the four known forces in the universe fit together (it doesn't account for gravity).

According to this theory, many fundamental particles had no mass immediately following the Big Bang, but gained mass later from interacting with an invisible energy field called the Higgs field, by way of a particle called the Higgs boson.

The Higgs boson is one of several missing pieces that make the Standard Model incomplete. Finding it with the Large Hadron Collider would lend additional credence to the Standard Model, giving us a strong indication of the nature of matter. Not finding it, after extensive searching, would indicate this theory is wrong, spurring physicists to focus on other schools of thought.

Professor Peter Higgs proposed his boson theory back in 1964

What exactly goes on in the LHC?

Some 50-170 metres (165-560 feet) beneath Switzerland and France, you'll find the Large Hadron Collider, a 27-kilometre (17-mile) circular racing track for atoms and subatomic particles. The collider accelerates and crashes streams of these particles into each other at 99.9999991 per cent the speed of light, in order to break them apart. So, why bother? Well, it takes collisions of this unprecedented intensity to get a look at some of the infinitesimally small particles that make up atoms. Examining these pieces is as close as physicists can get to seeing what the universe was like immediately after the Big Bang.

Researchers at the European Organisation for Nuclear Research (CERN) are collecting data on the speed, mass, energy, position, charge and trajectory of the particles in each collision. Analysis of the data could lead to new understandings of the nature of mass, gravity, dark matter and even other dimensions.

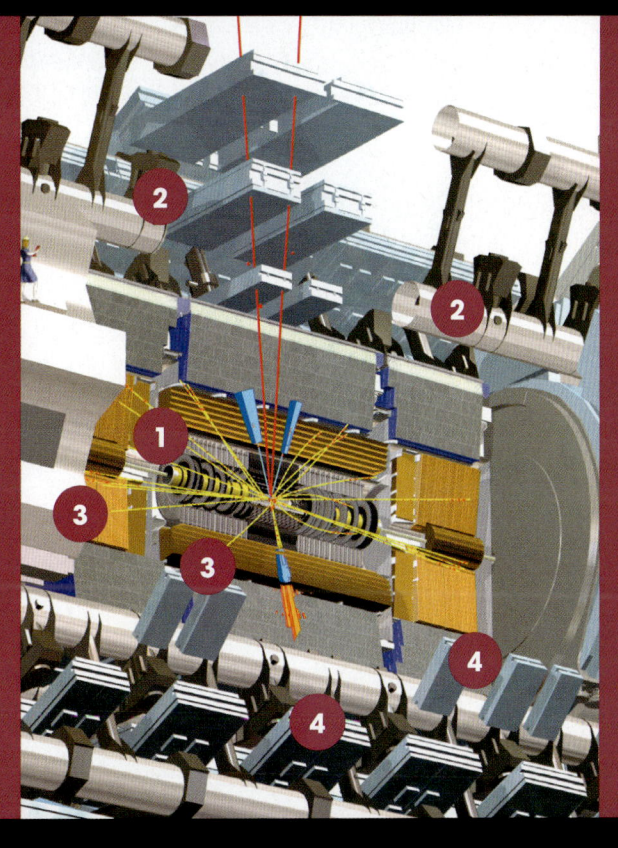

1 Inner detector
The innermost detector tracks the path of particles from proton collisions as they interact with various materials.

2 Toroid Magnet System
Powerful magnets bend the path of charged particles, making it possible to measure their momentum.

3 Calorimeter
A range of materials and sensors here absorb particles from collisions to measure their energy.

4 Muon spectrometer
The outermost detector has thousands of charged sensors that measure the momentum of muons, negatively charged subatomic particles.

UNDERSTANDING PHYSICS

The universe

112 Laws of the universe
Learn about the amazingly universal physics that governs the history, present and future of the cosmos

118 Why does Earth spin?
Find out why our planet has rotational movement

120 The secrets of lightspeed
Can we ever overcome the universe's ultimate speed limit?

126 Measuring a galaxy's mass
How we work out the weight of galaxies in the universe

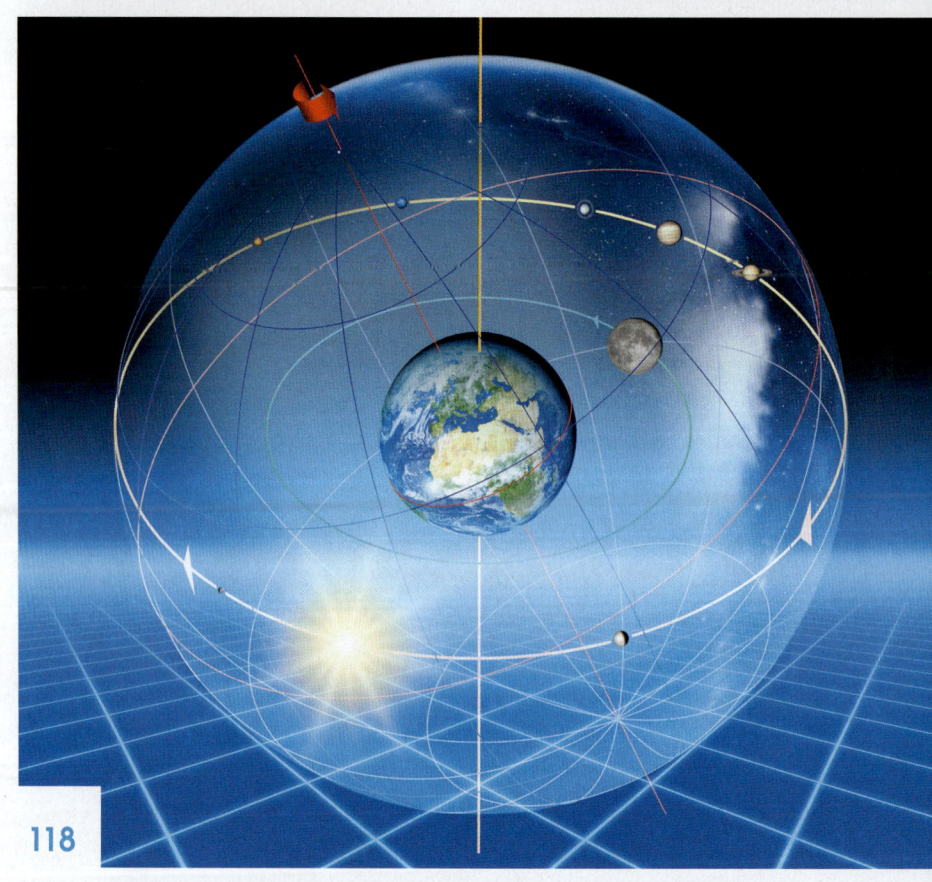

THE UNIVERSE

120

126

UNDERSTANDING PHYSICS

Laws of the universe

Learn about the amazingly universal physics which governs the history, present and future of the cosmos

From the intricate behaviour of subatomic particles to the gravitational dance of the largest galaxy clusters, our universe displays amazing, infinite complexity. Yet at heart it relies on just a small number of fundamental laws. Evidence points to just four forces – gravitation, electromagnetism, plus the weak and strong nuclear forces – governing every kind of interaction between matter. In effect, the large-scale universe is even simpler than that – the nuclear forces, as their name suggests, only make their influence felt over the tiny distances within atoms, while the infinite range of electromagnetism and gravity arguably make them the dominant powers.

All these laws, however, are required to make the cosmos behave in the way that we observe, and there's a good chance that if they, or the 'universal constants' that govern their influence, were just slightly different, the universe as we know it might not be here at all.

The laws that govern the universe have been gradually uncovered over several centuries. The effects of gravitation were first noted academically in the early-17th century, and the more fundamental laws that govern them towards the end of that century. The 19th century brought an improved understanding of energy and electromagnetism, while the 20th century revealed the quantum laws that govern atoms themselves, and transformed our understanding of gravity once again.

THE UNIVERSE

LAWS OF THE UNIVERSE

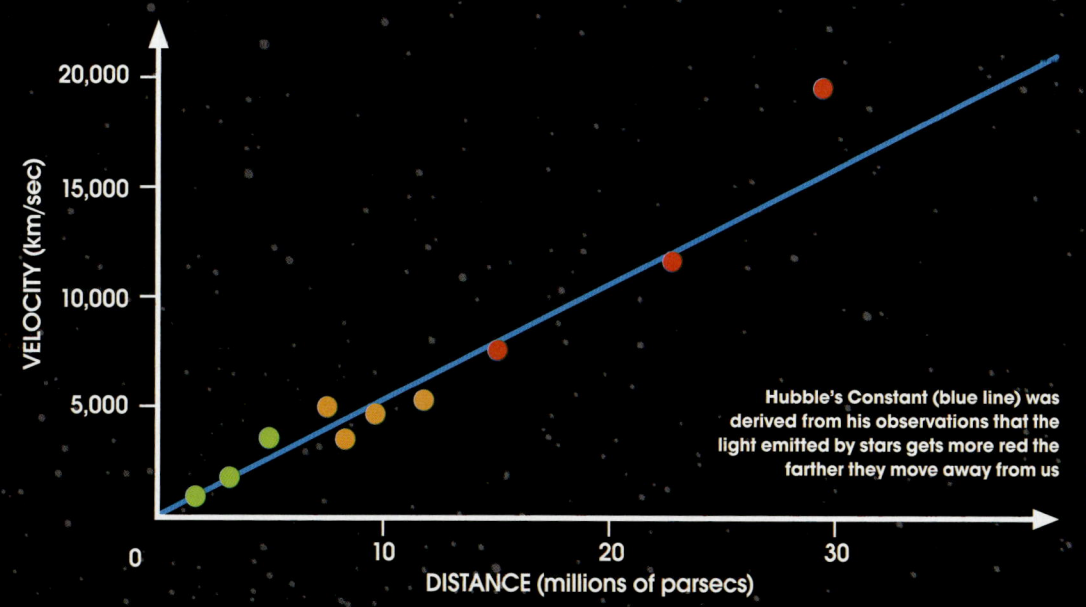

Cosmic expansion

In the mid-Twenties, astronomer Edwin Hubble used the behaviour of distant flickering variable stars to prove the 'spiral nebulas' in the sky were, in fact, galaxies millions of light years beyond our own. He went on to make an even more fundamental discovery. Light from many spiral nebulas was already known to be 'red shifted' – elongated in wavelength and shifted towards the red end of the spectrum due to the same Doppler effect that affects the pitch of a passing emergency siren. Hubble found that the farther away a distant galaxy was, the greater the red shift in its light. The only way to explain this was if the universe itself was expanding rapidly and dragging the galaxies apart.

Hubble's Constant (blue line) was derived from his observations that the light emitted by stars gets more red the farther they move away from us

Big Bang theory

Evidence including the expansion of the cosmos and an omnipresent 'afterglow' of radiation shows that the universe was created in an enormous explosion some 13.8 billion years ago. In the first moments of creation, the concentration of energy was so intense that the universe's fundamental forces acted as a single unified force – and indeed many cosmologists suspect that the separation of the forces helped drive a violent period of expansion known as inflation. Since that time, the cosmos has steadily expanded and cooled. For the first 380,000 years, the universe remained so dense and opaque that it was largely governed by electromagnetic interactions as light bounced back and forth between particles. Thereafter, the universe rapidly became transparent, and since then gravity has been the key factor that has shaped the cosmos.

KEY PLAYER
Edwin Hubble
US astronomer Edwin Hubble (1889-1953) used Cepheid variable stars to measure the distance to remote galaxies and prove they lay far beyond the Milky Way. He also devised a scheme for galaxy classification, and showed that the universe as a whole is expanding, paving the way for the Big Bang theory.

Evolution of the cosmos
Various forces and laws have made their influence felt at different points in the universe's long history...

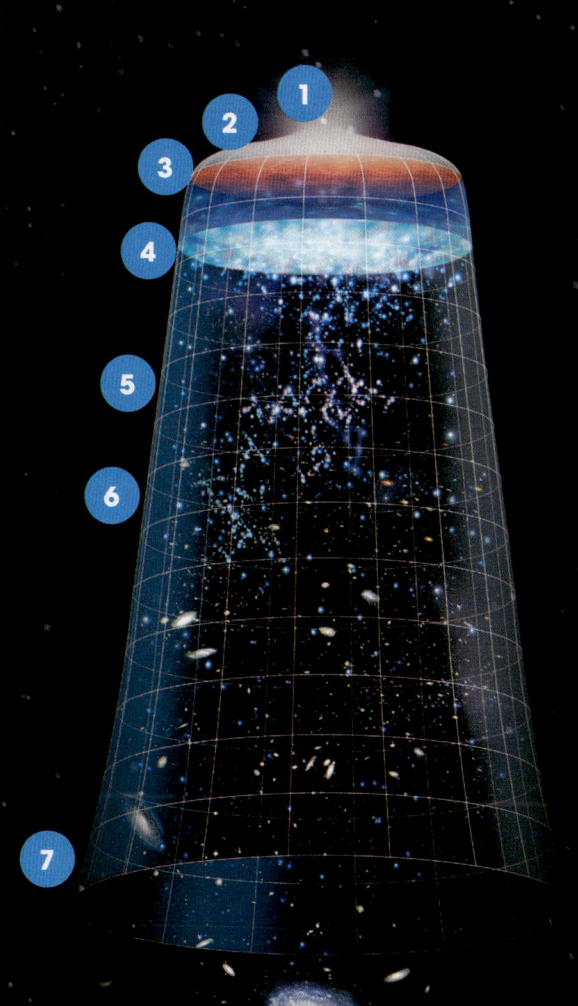

1 Big Bang
The universe sprang into existence in a cataclysmic explosion some 13.8 billion years ago.

2 Matter and energy
In the intensely hot conditions of the new universe, mass and energy were interchangeable, with particles of matter continually popping in and out of existence.

3 Inflation
Shortly after the initial explosion, separation of the four fundamental forces drove a brief period of sudden expansion and cooling known as inflation.

4 Radiation era
For some 380,000 years, the universe was opaque, and electromagnetic interactions governed the behaviour of visible matter.

5 Matter era
Once the cosmos became transparent, matter began to collapse as a result of gravity.

6 First stars
After around 300 million years, the first stars began to develop from collapsing gas clouds, transforming the lightweight elements created in the Big Bang into heavier ones.

7 The universe today
Today, the expansion of the universe continues on a grand scale, and is even accelerating thanks to the influence of mysterious dark energy.

UNDERSTANDING PHYSICS

Kepler's laws decoded

Three laws of planetary motion describe the movement of planets or any object in orbit around another under the influence of gravity

The first great physical laws to be discovered were those governing planetary motion. In the early-16th century, Nicolaus Copernicus had been the first modern astronomer to suggest the planets orbited the Sun, and evidence to support this radical idea had mounted through the century. But one of the essential jobs for any good theory of the planets was that it should predict their motion and positions, and this was where the Copernican theory fell down, offering little more accuracy than the old Earth-centred theory of the universe.

In 1609, German astronomer Johannes Kepler made a daring conceptual leap. Earlier generations of stargazers had been wedded to the idea of 'perfect' circular motion, but Kepler suggested that instead, the planets followed elliptical paths – ovals stretched along one axis, with the Sun at one of two focus points.

The resulting laws of planetary motion proved successful, but the underlying force behind them wasn't described until 1687, when Isaac Newton published his Mathematical Principles Of Natural Philosophy. Here, Newton showed that planetary orbits were just one manifestation of more fundamental laws of motion: that an object will continue in a state of rest or motion in a straight line unless acted on by a force; that the acceleration (a) experienced by a body of mass (m) under the influence of a force (F) is given by the simple equation a = F/m; and that when one body exerts a force on a second, it experiences an equal and opposite force. Newton argued that even when objects were not in physical contact, they could influence each other through gravity. He argued that the same force which causes objects to fall to the ground on Earth extends into space in accordance with universal gravitation, and that its influence can be infinite, making gravity the governing force shaping the large-scale universe.

KEY
- 🔴 **Law 1:** Law of ellipses
- 🔵 **Law 2:** Law of equal areas
- 🟢 **Law 3:** Law of harmonies

Shorter orbit
Objects close to the Sun have short periods not only because their orbits are themselves shorter, but also because they move more rapidly along them.

Solar focus
The object being orbited – in this case the Sun – sits at one of two focus points along the ellipse's long 'major' axis.

Perihelion
When the planet is at its closest point to the Sun, it moves fast and sweeps out a broad, short triangle.

Elliptical orbit
According to Kepler's first law, one object in orbit around another follows an elliptical path. Circular orbits are just a special type of ellipse.

Equal areas rule
Kepler's second law states that a triangle connecting points on the orbit to the focus sweeps out equal areas in equal times.

KEY PLAYER
Johannes Kepler
Kepler (1571-1630) was a student of the astronomer Tycho Brahe, and used his observations of planetary motions to form his revolutionary theory of elliptical planetary orbits. Despite being remembered as a scientific pioneer, he was also a keen astrologer.

From planets to galaxies...

Newton's laws of motion and gravitation can extend beyond our Solar System to describe the structure of galaxies, and even their large-scale movement. In a spiral galaxy, stars orbiting in the outer disc are affected by the same rules as planets, so in general they move more slowly at greater distances from the centre – a phenomenon called differential rotation that means galaxies don't spin like solid objects. Galaxies are also affected by the gravity of their neighbours, often concentrating in loose groups or denser clusters. Sometimes, however, galactic rotation and cluster dynamics fail to match the behaviour we might expect from visible matter, and this is a major clue indicating the existence of unseen 'dark matter' that is both invisible and transparent, making its presence felt only through its gravity.

THE UNIVERSE

LAWS OF THE UNIVERSE

Length of orbits
The orbital period (T) of an object is related to its semi-major axis (a) – that is, half the length of its long axis – by the equation $T^2 \propto a^3$.

Aphelion
At its farthest point from the Sun, the planet moves more slowly, sweeping out a long, narrow triangle.

Rule of distances
The total length of a line connecting one focus through the orbiting planet to the other remains constant at every point on the orbit.

Longer orbit
Objects at a greater distance have longer orbital periods not only because their orbits are longer, but also because they move more slowly through space.

KEY PLAYER
Isaac Newton
Mathematician and physicist Isaac Newton (1643-1727) devised the laws of motion and universal gravitation and the maths of calculus. He also contributed to the study of optics, building the first practical reflecting telescope and showing that white light consisted of many colours.

Thermodynamics: the rules of heat and energy

The science of thermodynamics studies the properties of heat and how it can be transferred. Temperature is, in essence, the random motion of particles within matter, and heat is the flow of energy by which this motion is transferred from one object to another. So stellar interiors can be modelled in terms of the different processes – conduction, convection and radiation – that transfer energy through them. Several important laws, largely developed in the 19th century, define the behaviour of thermodynamic systems, but the most important for cosmologists is the second law, which states that the entropy of a closed system (a measure of its disorganisation and the way heat energy is evenly spread across it) inevitably increases. Since in thermodynamic terms the universe is a closed system, this means that its entropy too will inevitably increase; in other words, heat will eventually become evenly spread out, dooming the cosmos to a long, cold death unless other factors intervene beforehand.

> "Newton showed that planetary orbits were just one manifestation of more fundamental laws of motion"

UNDERSTANDING PHYSICS

While the 18th and 19th centuries saw important developments in the theories of heat and energy (like thermodynamics), the early-20th century saw twin revolutions in the science of the very large and the very small. Albert Einstein's general theory of relativity, published in 1915, recast gravity as a distortion in the fabric of a four-dimensional 'space-time', created by large concentrations of mass.

It described and predicted phenomena that couldn't be accounted for by Newtonian gravity alone, such as gravitational lensing (the way that beams of light, which have no mass and therefore should be immune to the influence of gravity, are deflected when they pass close to massive objects). Meanwhile, Edwin Hubble's key observations proved that the universe as a whole is expanding, pointing to a much denser, hotter origin in the distant past: the Big Bang.

Einstein also played a key role in the formation of quantum theory – the idea that on the smallest scales, all phenomena display simultaneously wave-like and particle-like characteristics, and that matter and energy are interchangeable. First posed by Max Planck in 1900, and subsequently developed in the Twenties and Thirties by figures such as Louis de Broglie, Niels Bohr and Werner Heisenberg, quantum physics has provided the key to unlocking both the nature of light and other electromagnetic radiation and the structure of matter itself. The unpredictable nature of quantum-scale system helps to explain phenomena such as radioactive decay, but also poses some troubling philosophical questions.

Building on the success of quantum theory, physics in the late-20th century got to grips with the way forces work on the quantum scale, successfully developing 'gauge theories' that show how electromagnetism and the weak and strong nuclear forces are transmitted through the exchange of messenger particles called bosons between susceptible particles of matter. Based on this Standard Model of particle physics, and ideas such as the equivalence of mass and energy (embodied in Einstein's famous equation $E=mc^2$), cosmologists have been able to show how the energy released in the Big Bang could give rise to the raw materials that make up the universe.

Today, theoretical physicists are largely concerned with the pursuit to unify the fundamental forces of the cosmos in a simplified single model. But although these efforts have shown some positive results, a 'Theory of Everything' which describes all the laws of the universe in a neat unified equation remain some way off realisation.

KEY PLAYER

Albert Einstein

Physicist Albert Einstein (1879-1955) shaped 20th-century physics more than any other. His special and general theories of relativity revealed the interchangeability of mass and energy and the true natures of space, time and gravity. He also played a major role in developing quantum theory.

The nature of space-time
According to general relativity, our perceptions of space and time are aspects of a single four-dimensional space-time...

Pinched or warped?
In three dimensions, we might imagine that space-time is, in fact, 'pinched in' around large masses, a little like the waist of an hourglass.

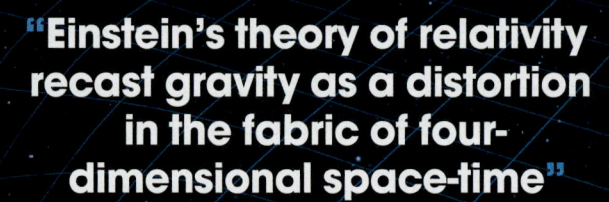

Gravitational well
Distorted space-time around the larger object creates a 'well'. Objects in orbit roll around the edge of this well at a distance determined by their speed.

Flat space
Space-time is only flat and uniform in regions that are far away from any mass or gravitational influence.

Distorting mass
Large masses such as the Earth distort space-time around them, warping it in a way that we perceive as gravity.

Depicting space-time
One common way of visualising the true nature of space-time is to depict space as a two-dimensional 'rubber sheet'.

Smaller distortion
Smaller masses such as the Moon create smaller dents in the 'sheet' of space-time.

> "Einstein's theory of relativity recast gravity as a distortion in the fabric of four-dimensional space-time"

THE UNIVERSE

LAWS OF THE UNIVERSE

A focus on fundamental forces

Four fundamental forces shape the universe: gravitation, electromagnetism and weak and strong nuclear forces. Electromagnetism is probably the simplest to understand; it acts on any object with an electric charge, felt as a force of attraction or repulsion. It transmits between objects by force-carrying particles called photons and 'virtual' photons.

The two nuclear forces are confined within the very heart of atoms, where they bind subatomic particles together. The strong force acts on particles called quarks, binding them in groups of three (through force-carrying gluon particles) to create hadrons such as protons and neutrons, and binding these particles together in turn to make complete atomic nuclei. The weak force, meanwhile, is carried by two different types of particle – known as the W and Z bosons – and this can trigger the spontaneous changes seen in the nuclei of atoms that are linked to radioactive decay.

Gravitation is the most mysterious force of all – much weaker than the others, but influential over an infinite range. Although general relativity describes it as a distortion of space-time, some scientists still speculate that it may be transmitted by theorised particles referred to as gravitons. In recent decades, physicists have begun to uncover deep connections between these apparently different forces. At high enough energies, electromagnetism and the weak force merge into a single electroweak force, and there are hopes that at even higher energies the strong force may eventually be combined with them to create a so-called 'Grand Unified Theory'.

The Big Bang

Theoretical unified force

Electroweak force

Weak **Electromagnetic** **Strong** **Gravitational**

Challenging universal laws

While many believe the laws of physics work in the same way throughout the universe, there's evidence that their effects might vary. Equations like Newton's laws of universal gravitation contain elements called constants, which determine, for example, the strength of gravitational force exerted at a certain distance. But a 2010 survey has found evidence that these so-called 'constants' are nothing of the kind, hinting that the 'fine structure' constant might be different from one part of the universe to another, dictating the strength of the electromagnetic force. Similarly, recent measurements suggest the strength of the universal gravitational constant may also be changing, weakening gravity. While the fluctuations we can see are tiny, it's possible the constants vary more elsewhere.

Data from the Keck Observatory has cast doubt on what was believed to be a natural constant

Parallel possibilities?

Parallel universes may sound like science fiction, but the laws of physics as we understand them permit them to exist in various ways. One possibility is that the universe is infinite and stretches beyond the region of space we can see, in which case we could expect every possible set of conditions and 'universe' to occur somewhere, and might even be able to reach some of them via theoretical shortcuts called wormholes (illustrated below). Another theory is that our cosmos is one of many four-dimensional 'branes' floating like sheets in multidimensional space, with parallel universes existing on similar sheets that may be out of reach.

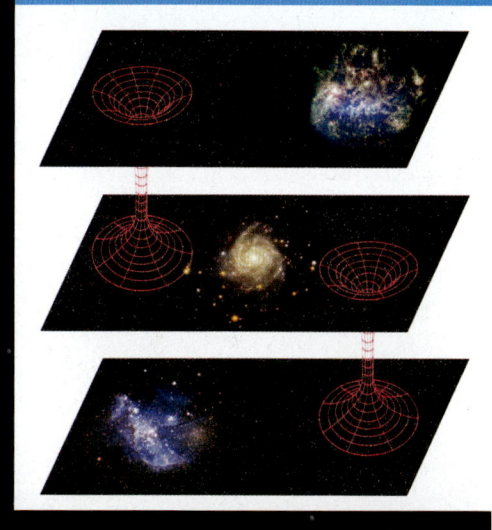

KEY PLAYER
Max Planck

Physicist Max Planck (1858-1947) is widely recognised as the founder of quantum theory thanks to his suggestion that light and other forms of electromagnetic radiation might be found in discrete packets of energy, or quanta. He argued against the wider implications of quantum physics.

KEY PLAYER
Stephen Hawking

Physicist and cosmologist Stephen Hawking (born 1942) dedicated his early career to understanding the gravitational laws governing black holes. In more recent years, he has developed cosmological theories that offer hope for unifying the theories of general relativity and quantum theory.

117

UNDERSTANDING PHYSICS

Why does the Earth spin?

Find out why our planet has rotational movement

The story of why the Earth spins goes back to the formation of the solar system. In the beginning, roughly 4.7 billion years ago, the solar system was a large swirling cloud of dust and gas. Over time this gradually coalesced into stars and planets, being drawn into these shapes by gravity. This motion of being pulled inwards increased the angular momentum of the various bodies, and thus caused them to start rotating faster.

Consider an ice skater spinning with their arms outstretched. As they spin, they bring their arms inwards. Doing so increases their angular momentum, which makes them spin faster. The same is true for when the Earth first formed. As the dust and gas was compressed into one solid mass, the total mass of the object became more confined and subsequently began to rotate more and more rapidly.

The law of inertia states that anything stationary or moving with a constant speed wants to continue doing so until it is acted upon by another body/force. Considering the Earth rotates in space, which is a vacuum, there is nothing to drastically slow the Earth down which is why it continues to spin. Interestingly, early in its formation, the world spun up to five times faster than it does now – so we know Earth has lost some speed.

The culprit is the moon. It is our own natural satellite that has caused the planet to slow down, via something known as tidal locking. The moon at the moment is tidally locked to the Earth – that is, the same face always looks towards us, but it was not always so. When the moon first came into orbit around the Earth it was also spinning. To understand tidal locking, imagine that you and a friend both pull on a piece of rope, but at the same time you spin in a circle around a pivot at the centre of the rope. As you tug harder, you are eventually able to spin less and less fast. Eventually, you will be stuck simply pulling on the rope, unable to move sideways as your pulling force is too great; this is essentially what happened between Earth and the moon. As the moon orbits the world it exerts a pull on the planet, which is responsible for causing tides. The Earth is much bigger so it continues to spin freely, but the moon's rotation now matches the time it takes to complete one orbit.. Small as it may be, the moon will continue to have an effect on Earth and, millions of years from now, a day on Earth could be up to 26 hours long.

CELESTIAL EQUATOR

Inertia
The law of inertia states an object will continue to move unless acted upon by another force, which is why Earth has not stopped spinning.

Ecliptic
The Earth orbits the Sun on a flat plane but it does not rotate perpendicular to this plane.

Poles
The poles experience little to no rotational force, and thus can experience prolonged daylight and darkness in summer and winter, respectively.

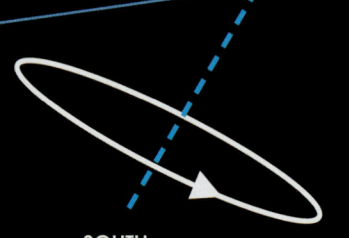

SOUTH CELESTIAL POLE

THE UNIVERSE

WHY DOES THE EARTH SPIN?

PERPENDICULAR TO ORBIT

NORTH CELESTIAL POLE

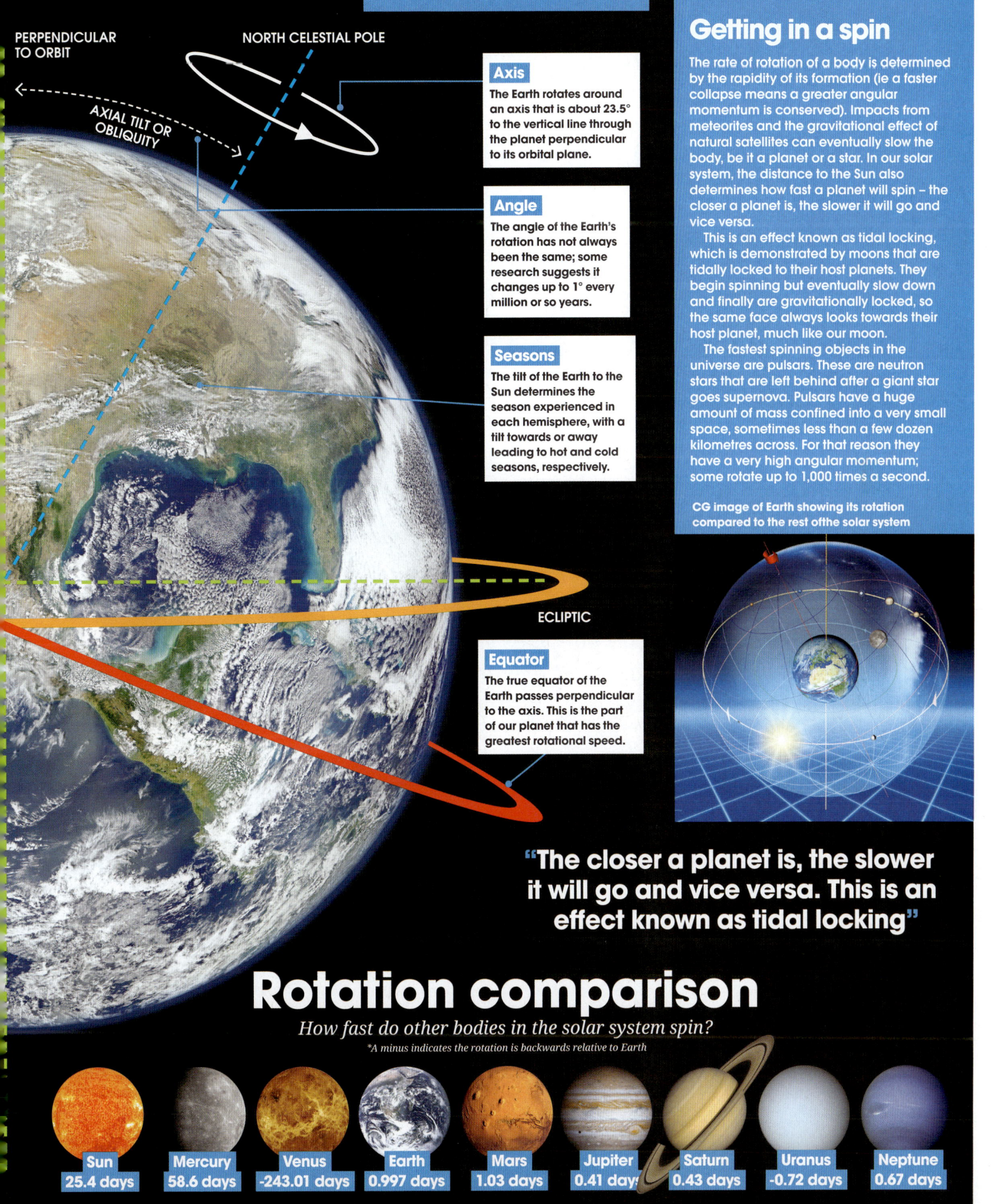

AXIAL TILT OR OBLIQUITY

Axis
The Earth rotates around an axis that is about 23.5° to the vertical line through the planet perpendicular to its orbital plane.

Angle
The angle of the Earth's rotation has not always been the same; some research suggests it changes up to 1° every million or so years.

Seasons
The tilt of the Earth to the Sun determines the season experienced in each hemisphere, with a tilt towards or away leading to hot and cold seasons, respectively.

ECLIPTIC

Equator
The true equator of the Earth passes perpendicular to the axis. This is the part of our planet that has the greatest rotational speed.

Getting in a spin

The rate of rotation of a body is determined by the rapidity of its formation (ie a faster collapse means a greater angular momentum is conserved). Impacts from meteorites and the gravitational effect of natural satellites can eventually slow the body, be it a planet or a star. In our solar system, the distance to the Sun also determines how fast a planet will spin – the closer a planet is, the slower it will go and vice versa.

This is an effect known as tidal locking, which is demonstrated by moons that are tidally locked to their host planets. They begin spinning but eventually slow down and finally are gravitationally locked, so the same face always looks towards their host planet, much like our moon.

The fastest spinning objects in the universe are pulsars. These are neutron stars that are left behind after a giant star goes supernova. Pulsars have a huge amount of mass confined into a very small space, sometimes less than a few dozen kilometres across. For that reason they have a very high angular momentum; some rotate up to 1,000 times a second.

CG image of Earth showing its rotation compared to the rest of the solar system

> "The closer a planet is, the slower it will go and vice versa. This is an effect known as tidal locking"

Rotation comparison
How fast do other bodies in the solar system spin?
*A minus indicates the rotation is backwards relative to Earth

Sun	Mercury	Venus	Earth	Mars	Jupiter	Saturn	Uranus	Neptune
25.4 days	58.6 days	-243.01 days	0.997 days	1.03 days	0.41 days	0.43 days	-0.72 days	0.67 days

UNDERSTANDING PHYSICS

The secrets of lightspeed
& the fastest phenomena in space

High-velocity particles can tell us a lot about the way the universe works – but can we ever overcome the ultimate speed limit?

For a few months in early-2012, the scientific world held its breath as researchers raced to establish whether one of the greatest tenets of modern physics was under threat. The panic was triggered by reports from the Gran Sasso National Laboratory, beneath Italy's Apennine Mountains, which appeared to show bursts of neutrinos (tiny, near-massless subatomic particles), fired from a particle accelerator at CERN on the Swiss/French border some 730 kilometres (454 miles) distant, travelling faster than the speed of light.

According to more than a century of established physics, the speed of light in a vacuum, 299,792.458 kilometres (186,282.397 miles) per second – is the ultimate speed limit of the universe. No object with mass can reach this speed for very good reasons outlined in the work of Albert Einstein; as they get close, travelling at so-called 'relativistic' speeds, the strange effects predicted by Einstein's theory of special relativity take effect, including time slowing down, distances contracting and mass increasing (making it ever-more difficult to accelerate). Only massless photons of light and other electromagnetic radiation can reach the speed of light itself.

Sadly for those anticipating a revolution in physics sources, rigorous checking at Gran Sasso eventually identified errors in the timing of the neutrino bursts, confirming they had, in fact, not exceeded the speed of light: for the moment at least, the status quo prevails.

But 'superfast' doesn't always have to threaten the fundamental laws of physics – objects moving far faster than we would expect, even if not at relativistic speeds, can still present us with intriguing puzzles.

Looked at from this perspective, our universe is full of superfast phenomena – from weird particles that get within a trillionth of a per cent of light speed itself, to planets, stars and even man-made space probes moving far, far faster than a speeding bullet.

THE UNIVERSE

THE SECRETS OF LIGHTSPEED

Special relativity and the ultimate speed limit

Albert Einstein developed his theory of special relativity in order to resolve a crisis in physics during the late-19th century. As methods for measuring the speed of light got more and more accurate, it became clear that it did not behave like other phenomena – its speed was always the same, regardless of the relative motions of source and observer. Physicists tried various tricks to get around the problem, but Einstein was the only person who dared to tackle it head on. He rewrote the laws of physics from the ground up based on two simple principles: a fixed speed of light and the 'principle of special relativity' – that the laws of physics should appear the same for all observers in 'inertial reference frames' (situations and viewpoints not involving acceleration or deceleration).

Einstein showed that objects moving at 'relativistic' speeds (superfast speeds comparable to that of light) must experience distortions in their apparent mass, length and even the flow of time (as seen from the point of view of an outside observer). These distortions become infinite when an object attempts to move at the speed of light itself, convincing Einstein that light speed is the ultimate speed limit. Einstein's theory now has more than a century of experimental observations to back it up.

A photo of Albert Einstein, the father of the theory of relativity, circa 1947

A jet of electrons and subatomic particles travelling at relativistic speeds, powered by a supermassive black hole at the heart of galaxy M87

UNDERSTANDING PHYSICS

Quick-fire blazar jets

Blazars are distant 'active galaxies' with the supermassive black holes in their cores feeding voraciously on matter from their surroundings. Gas and dust spiralling into the black hole forms a superhot disc that from a distance looks like a rapidly changing, starlike point of light, while a powerful magnetic field spits out jets of particles perpendicular to the disc at relativistic speeds. In other types of active galaxy, we see this jet at an angle, but in blazars, the axis of the jets points more or less straight toward Earth. This creates an illusion of faster-than-light motion – material moving along the jet is almost able to keep pace with the radiation it emits, so emissions from a knot of material emitted near the blazar's core arrive at Earth just shortly after those emitted by the same material much farther out, giving the impression that the knot may be moving at many times the speed of light, but this is an illusion.

Hunting for blazars

The first blazars to be discovered were initially thought to be unusual variable stars – it was only in 1968 that astronomers discovered that they emit radio waves and appeared to be embedded within faint elliptical 'host galaxies' – characteristics similar to quasars, another type of active galaxy nucleus (AGN). Today, astronomers estimate the distance to blazars by measuring the 'red shift' in light from the host galaxies – an indication of how fast they are moving away from us due to the overall expansion of the universe, and therefore how far away they are. By imaging individual radio-emitting blobs shooting out of the galaxy's central nucleus, they can then calculate both the apparent and true speed of the jets.

Head-on view
A blazar is an example of an active galactic nucleus (AGN) at the heart of a galaxy, but unlike a quasar which is side-on, it faces Earth head-on.

Disc torus
An accretion disc of dust and other space matter is pulled toward the heart by the intense gravity of a black hole at the centre of the AGN.

Relativistic jet
At the centre of the blazar two jets of gamma-ray radiation shoot out at near light speed – one toward and one away from us. The light can be more than 1 billion times more energetic than our eyes can see.

Can we break the light barrier?

Einstein's theory of special relativity makes a convincing case that matter cannot travel at the speed of light – but what about speeds beyond light speed? Inspired by 2012's reports of possible faster-than-light neutrinos, mathematicians Jim Hill and Barry Cox of the University of Adelaide took a fresh look at the equations of special relativity and reached some surprising conclusions. They found that the equations can be elegantly extended beyond light speed towards infinity, with properties that mirror those approaching light speed (for example, the mass of objects approaching infinite speed would decrease toward zero).

Their findings put long-standing ideas about faster-than-light particles known as tachyons on a mathematical footing, but Hill and Cox emphasise that their ideas are based in maths: "We're mathematicians, not physicists, so we're approaching the problem from a theoretical mathematical perspective," explains Cox. "Our paper doesn't explain how this could be achieved, just how equations of motion might operate in (faster-than-light) regimes."

What's more, the equations still break down at the speed of light itself (where they produce mathematical 'infinities' that cannot be used to make physical predictions) – so it seems making the ultimate leap to faster-than-light travel is still some way off.

"In blazars, the axis of the jets points more or less straight toward Earth"

THE UNIVERSE

THE SECRETS OF LIGHTSPEED

Star
The Hubble telescope spotted a speedy USPP passing in front of this red dwarf star. The planet is 1/130th the distance of Earth from the Sun.

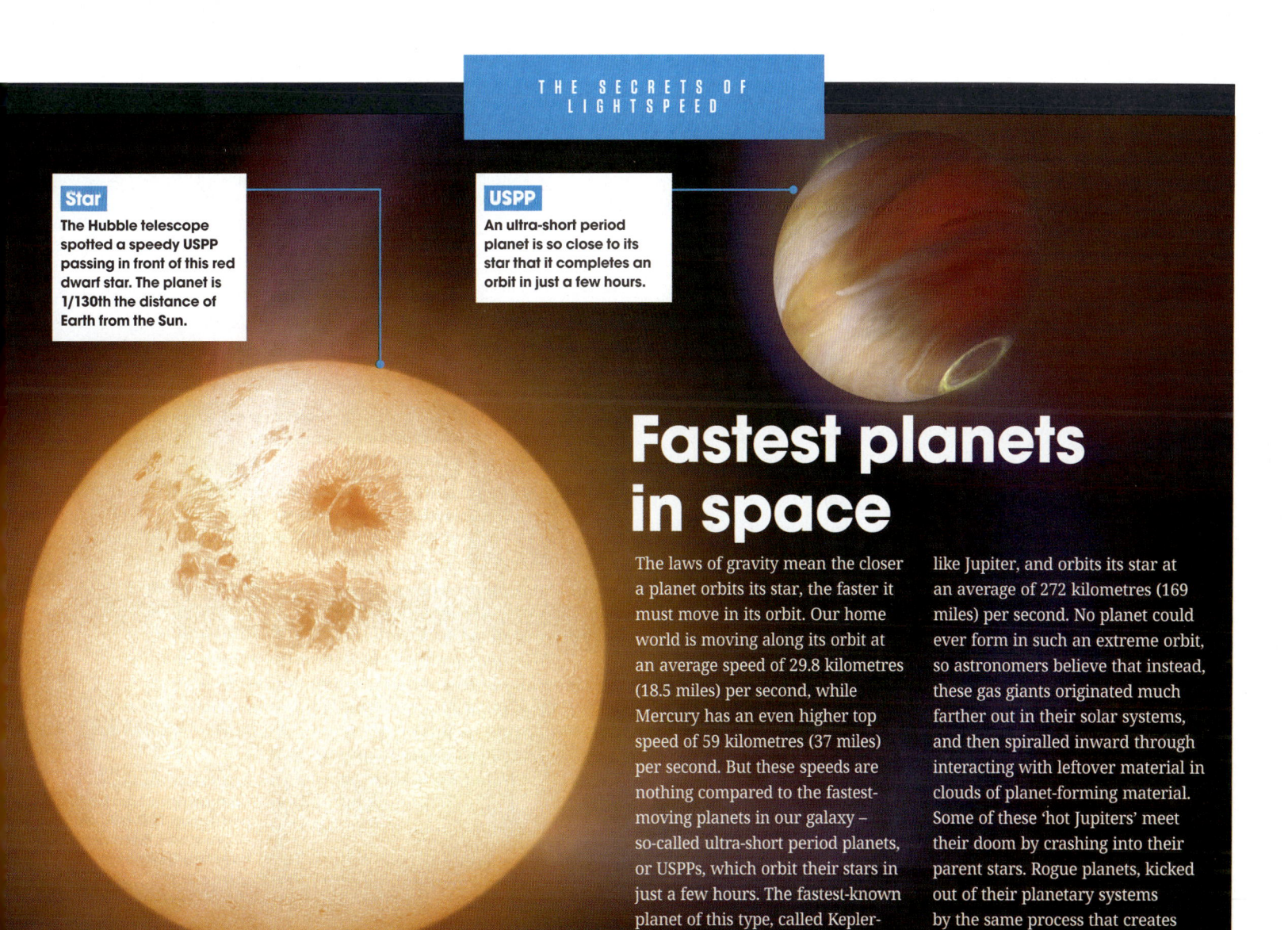

USPP
An ultra-short period planet is so close to its star that it completes an orbit in just a few hours.

Fastest planets in space

The laws of gravity mean the closer a planet orbits its star, the faster it must move in its orbit. Our home world is moving along its orbit at an average speed of 29.8 kilometres (18.5 miles) per second, while Mercury has an even higher top speed of 59 kilometres (37 miles) per second. But these speeds are nothing compared to the fastest-moving planets in our galaxy – so-called ultra-short period planets, or USPPs, which orbit their stars in just a few hours. The fastest-known planet of this type, called Kepler-70b, is thought to be the exposed solid core of a planet that was once like Jupiter, and orbits its star at an average of 272 kilometres (169 miles) per second. No planet could ever form in such an extreme orbit, so astronomers believe that instead, these gas giants originated much farther out in their solar systems, and then spiralled inward through interacting with leftover material in clouds of planet-forming material. Some of these 'hot Jupiters' meet their doom by crashing into their parent stars. Rogue planets, kicked out of their planetary systems by the same process that creates hypervelocity stars (see over page), can also achieve great speeds.

Cosmic rays: the fastest particles

Cosmic rays are particles moving at extremely high speed through space, originating from outside our Solar System. They rarely reach the surface of Earth intact, disintegrating into showers of lighter, lower-energy particles after colliding with gases in the upper atmosphere. Nevertheless, by tracking the speed and distribution of these secondary particles (and using satellite and balloon-based detectors), astronomers can discover a surprising amount about the properties of primary cosmic rays.

Mostly atomic nuclei of hydrogen and helium – the two lightest elements – with small amounts of heavier nuclei such as lithium and beryllium, they fall into two distinct categories. Most 'normal' cosmic rays travel at speeds of around 99 per cent of the speed of light. Trillions of them bombard Earth every single second and evidence suggests that a significant proportion of them were ejected from distant supernovas.

A much rarer population of ultra-high energy cosmic rays (UHECRs), meanwhile, carry far more energy and travel at speeds a tiny fraction of a per cent below light speed itself. UHECR sources seem to lie in the same direction as distant active galaxies, and some astronomers believe they are created by fast-spinning supermassive black holes acting as natural particle accelerators.

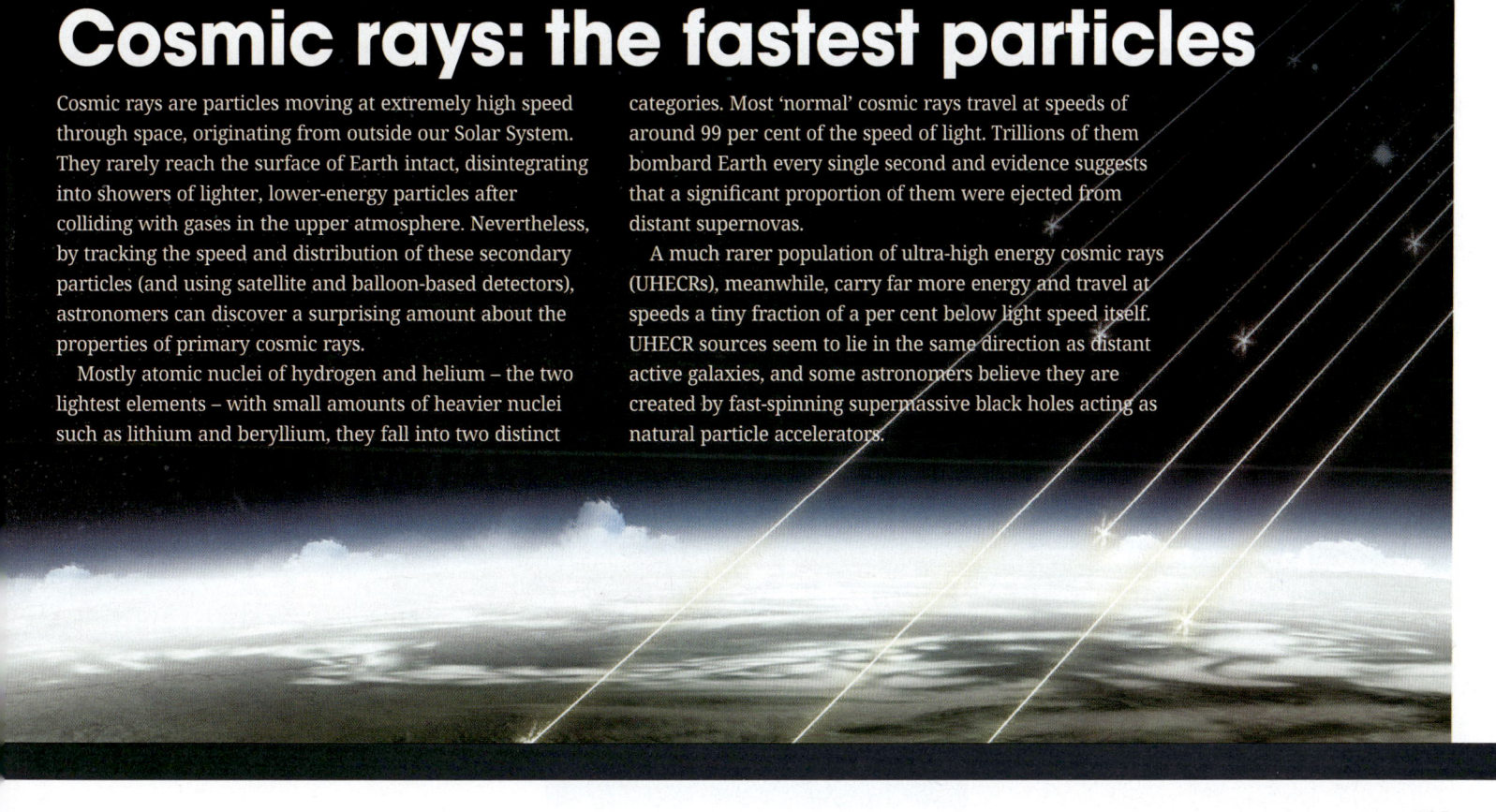

UNDERSTANDING PHYSICS

Quickest-ever spacecraft

In October 2013, the Jupiter-bound Juno spacecraft flew past Earth in a gravitational 'slingshot' manoeuvre that boosted its speed to become the fastest man-made object in the universe, shooting past us at nearly 40 kilometres (25 miles) per second relative to the Sun. Juno's slingshot made use of a technique that has been used on probes to distant planets since the 1970s, in which a spacecraft allows itself to be 'dragged in' by a planet's gravity field and accelerated, before swinging close to the planet and escaping along a different trajectory with a precisely timed burn of its rocket engines. The probe keeps the same speed relative to the planet's surface, but because the planet is moving, it can radically change its speed relative to the Solar System as a whole – in effect, the spacecraft steals a little of the planet's orbital momentum, but because the planet is so much heavier than the spacecraft, a little stolen momentum can have a dramatic effect.

Destination Jupiter
Juno's unique flight path to Jupiter will allow it to investigate unseen parts of the giant planet

Communications antenna
Juno's radio antenna doubles as a scientific instrument, allowing scientists to measure tiny variations in the spacecraft's speed caused by Jupiter's gravity field.

Slow spin
Juno spins on its axis once every 30 seconds, helping to keep its flight path stable.

Solar panels
Juno is the first mission to the outer Solar System to rely on solar panels for energy. Each is 2.7m (8.7ft) wide and 9m (29.5ft) long.

Magnetometer
This will measure Jupiter's powerful magnetic field in more detail than ever before.

Scientific payload
An array of instruments in the spacecraft's body will study Jupiter's atmosphere, magnetism and radiation as well as imaging the surface.

Stowaways
Juno also carries three tiny Lego figures, representing the Roman god Jupiter, his wife Juno and the Italian scientist Galileo.

Earth departure
Juno launched from Earth on 5 August 2011 onto an elliptical orbit that reached some way beyond Mars.

Deep-space manoeuvres
Two course corrections in August and September 2012 set Juno on course for its Earth flyby.

Jupiter rendezvous
The spacecraft is due to arrive at Jupiter in July 2016.

Earth flyby
Juno swung back past Earth in October 2013, picking up a huge speed boost that flung it on a final trajectory toward Jupiter.

> "In effect, the spacecraft steals a little of the planet's orbital momentum"

THE UNIVERSE

THE SECRETS OF LIGHTSPEED

Hypervelocity stars

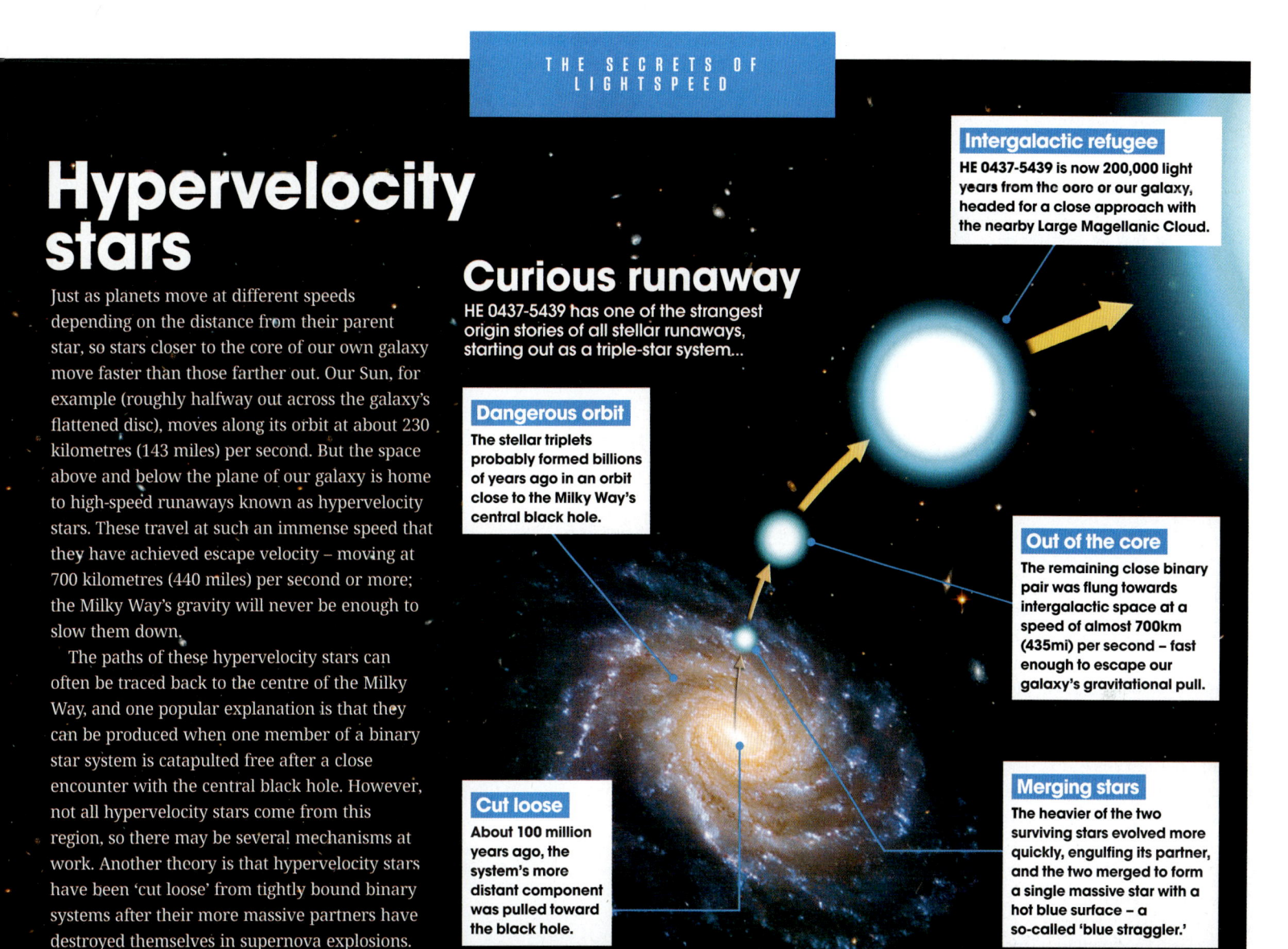

Just as planets move at different speeds depending on the distance from their parent star, so stars closer to the core of our own galaxy move faster than those farther out. Our Sun, for example (roughly halfway out across the galaxy's flattened disc), moves along its orbit at about 230 kilometres (143 miles) per second. But the space above and below the plane of our galaxy is home to high-speed runaways known as hypervelocity stars. These travel at such an immense speed that they have achieved escape velocity – moving at 700 kilometres (440 miles) per second or more; the Milky Way's gravity will never be enough to slow them down.

The paths of these hypervelocity stars can often be traced back to the centre of the Milky Way, and one popular explanation is that they can be produced when one member of a binary star system is catapulted free after a close encounter with the central black hole. However, not all hypervelocity stars come from this region, so there may be several mechanisms at work. Another theory is that hypervelocity stars have been 'cut loose' from tightly bound binary systems after their more massive partners have destroyed themselves in supernova explosions.

Curious runaway
HE 0437-5439 has one of the strangest origin stories of all stellar runaways, starting out as a triple-star system...

Dangerous orbit
The stellar triplets probably formed billions of years ago in an orbit close to the Milky Way's central black hole.

Cut loose
About 100 million years ago, the system's more distant component was pulled toward the black hole.

Intergalactic refugee
HE 0437-5439 is now 200,000 light years from the core of our galaxy, headed for a close approach with the nearby Large Magellanic Cloud.

Out of the core
The remaining close binary pair was flung towards intergalactic space at a speed of almost 700km (435mi) per second – fast enough to escape our galaxy's gravitational pull.

Merging stars
The heavier of the two surviving stars evolved more quickly, engulfing its partner, and the two merged to form a single massive star with a hot blue surface – a so-called 'blue straggler.'

Warp factor: fact or fiction?

Einstein's theory of special relativity suggests that it is impossible to move across space faster than the speed of light (or at least, to pass through the light-speed barrier), but could future space pioneers find ways to overcome this problem? One option would be to make use of the time dilation effect; time would flow more slowly for the crew on board a spacecraft that is moving at relativistic speeds, perhaps allowing them to travel across hundreds of light years in what, for those on the ship, would seem like only a few months.

But Einstein's general theory of relativity, which demonstrates that space-time is a four-dimensional 'manifold' that can be warped and distorted, offers another alternative – the 'warp drive'. First outlined in 1994 by Mexican physicist Miguel Alcubierre, such a device would involve moving a 'bubble' of normal space across great distances by compressing the region of space-time ahead of it and expanding the region behind it. A spacecraft inside the bubble could move at normal speeds relative to its immediate surroundings, while the bubble itself could move at faster-than-light speeds without actually breaking Einstein's rules.

NASA scientist Harold 'Sonny' White has since shown a doughnut-shaped region of distorted space-time could radically reduce the energy needs of a warp drive, and although the practical challenges remain huge, White's team at the Johnson Space Center have begun experiments to demonstrate warp effects at a micro level, which might one day be upscaled. So there's still hope for a real-life Starship Enterprise yet.

UNDERSTANDING PHYSICS

Measuring a galaxy's mass

How we work out the weight of galaxies in the universe

The idea of weighing a galaxy might seem a bit weird – it's not like we can put it on a giant cosmic set of scales. But by using some clever observations and tricky equations, it is indeed possible to work out the masses of other galaxies in the universe.

There are actually a number of ways to do it, but one popular method is to look at the orbital motion of stars in a galaxy. Those in a more massive galaxy will move faster than those in a less massive one, so measuring their speed can help scientists work out the answer. Scientists also look at the overall rotation rate of galaxies to work out their mass. They do this by measuring the redshift or blueshift – the amount that a particular side of a galaxy is moving away from or towards us respectively – and seeing how much the light shifts to each end of the spectrum. Another method involves looking at the gravitational pull exerted on star clusters in space by nearby galaxies. The bigger the pull, the more massive the galaxy, and we can use this to estimate just how heavy it really is.

Yet another method involves gravitational lensing, which is the lensing effect caused when a galaxy passes in front of a distant object in our line of sight. Depending on the gravitational strength (and therefore mass) of the lensing galaxy, this can produce either a large or small lensing effect, something that was predicted by Einstein and that is known as an Einstein ring. However, these events are rare in the universe, so the chances of us measuring a galaxy in this way are slim.

An Einstein ring can help us measure the mass of a lensing galaxy

Two measuring methods

Here are two ways used to estimate the mass of galaxy ESO 325-G004

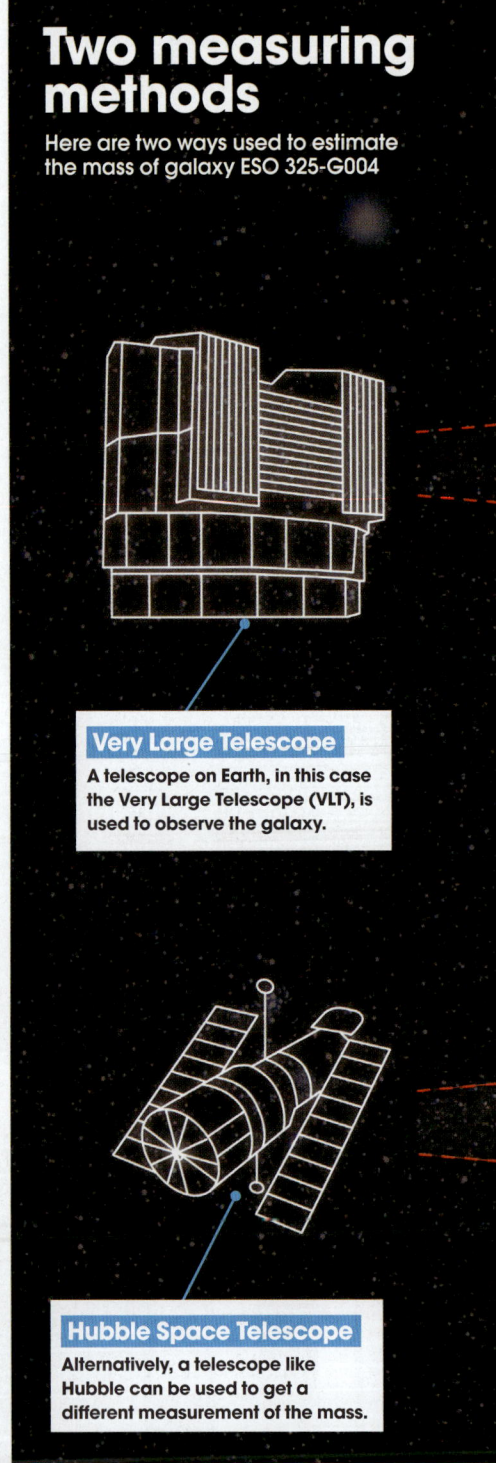

Very Large Telescope
A telescope on Earth, in this case the Very Large Telescope (VLT), is used to observe the galaxy.

Hubble Space Telescope
Alternatively, a telescope like Hubble can be used to get a different measurement of the mass.

The Milky Way

We can measure the mass of other galaxies, but how do we measure our own when we can't see the big picture from afar? The best way to do it is to measure the speed and motion of stars in our galaxy, although to get an accurate reading you need to measure at least 100,000 stars, and maybe millions to be sure.

Another way to get a good estimate is to look at 'tracers' behind our galaxy. These are rogue stars and galaxy fragments that now trail behind us, and their velocity and angular momentum can tell us how much they have been pulled by our galaxy, thus indicating its mass. The most recent estimate suggests our galaxy is about 960 billion times the mass of our Sun.

As we're inside the Milky Way it can be tricky to measure its mass

THE UNIVERSE

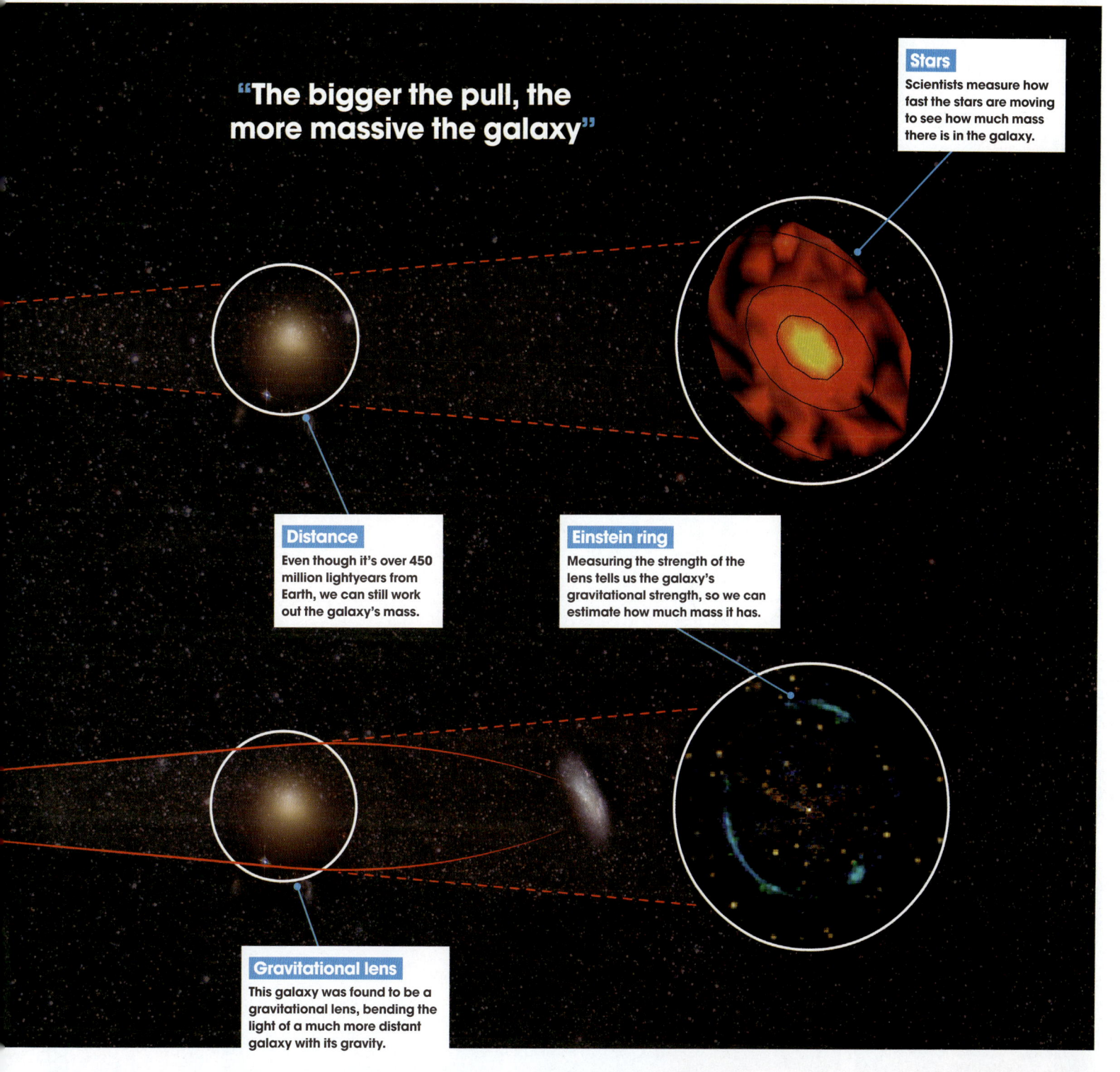

MEASURING A GALAXY'S MASS

"The bigger the pull, the more massive the galaxy"

Stars
Scientists measure how fast the stars are moving to see how much mass there is in the galaxy.

Distance
Even though it's over 450 million lightyears from Earth, we can still work out the galaxy's mass.

Einstein ring
Measuring the strength of the lens tells us the galaxy's gravitational strength, so we can estimate how much mass it has.

Gravitational lens
This galaxy was found to be a gravitational lens, bending the light of a much more distant galaxy with its gravity.

Hidden mass

One issue with measuring the mass of galaxies is the galaxy rotation curve problem. When an ice skater pulls in their arms during a spin, you would expect them to move faster as their mass is more concentrated towards their centre. So in a galaxy, you'd also expect the stars nearer the centre to move faster, but that's not the case. Instead we find that stars towards the edge of a rotating galaxy actually move faster, not slower. Scientists think the answer is the presence of invisible dark matter. The gravitational tug of dark matter causes the stars at the edge to move faster than theory predicts and thereby helps iron out any discrepancies in the measurement of a galaxy's gravity and mass.

Dark matter is thought to affect the rotation rate of galaxies

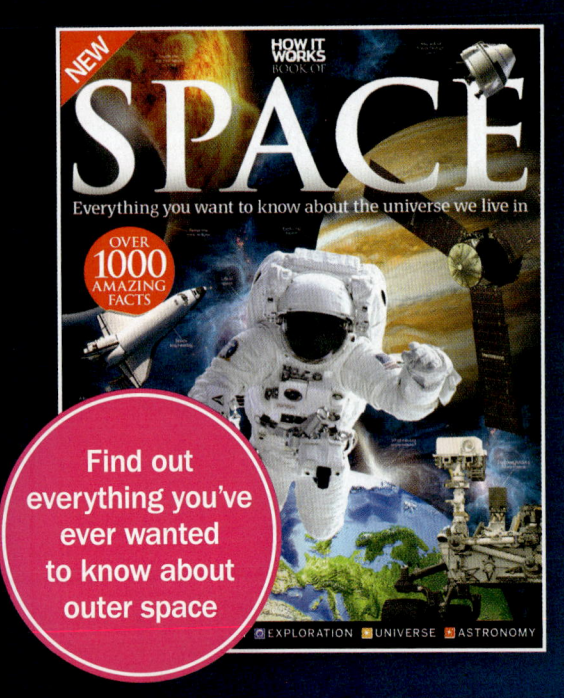

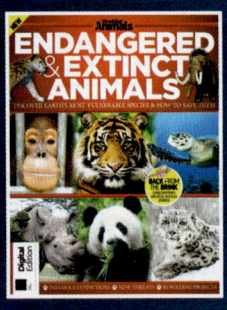

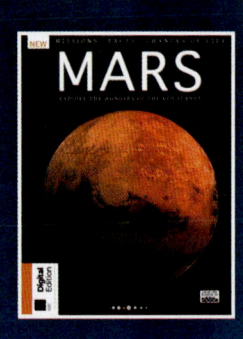

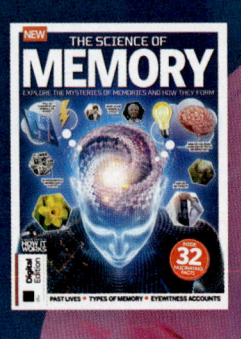

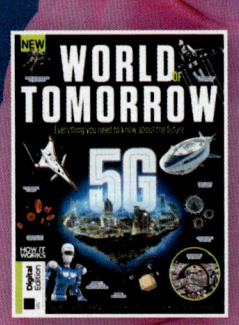

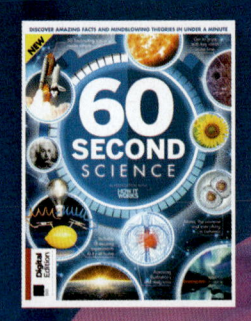

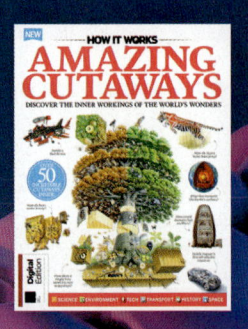

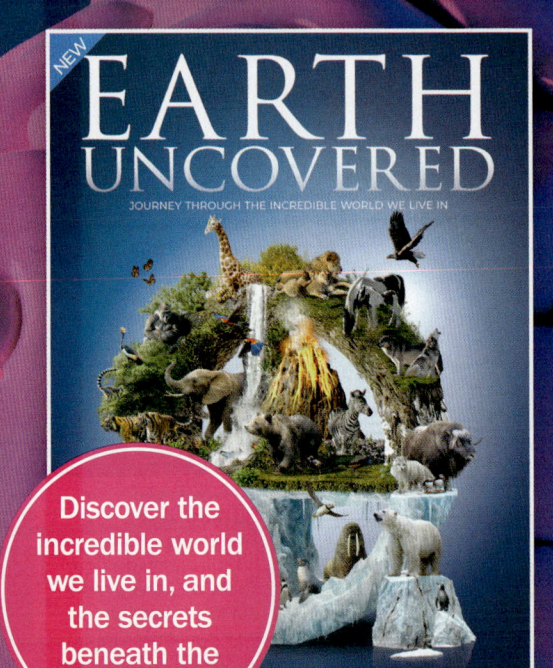

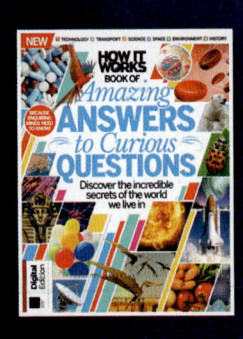

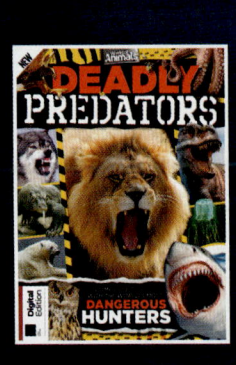

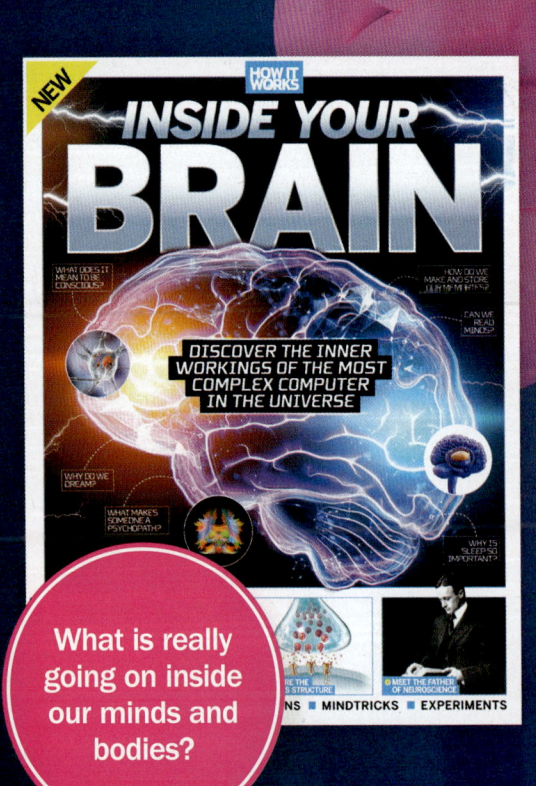

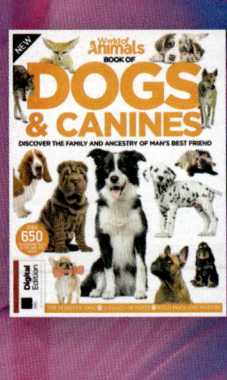

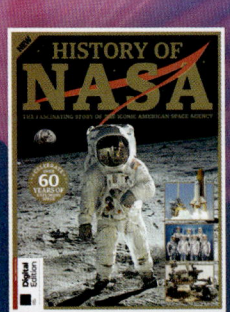

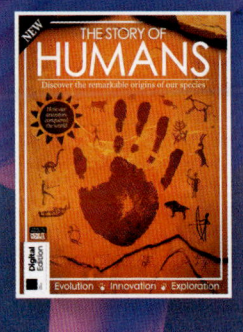

Find out everything you've ever wanted to know about outer space

Discover the incredible world we live in, and the secrets beneath the surface

What is really going on inside our minds and bodies?

✓ Get great savings when you buy direct from us

✓ 1000s of great titles, many not available anywhere else

✓ World-wide delivery and super-safe ordering

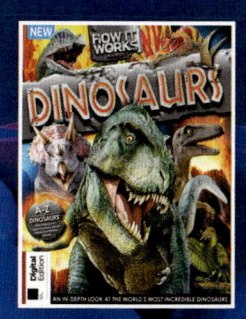

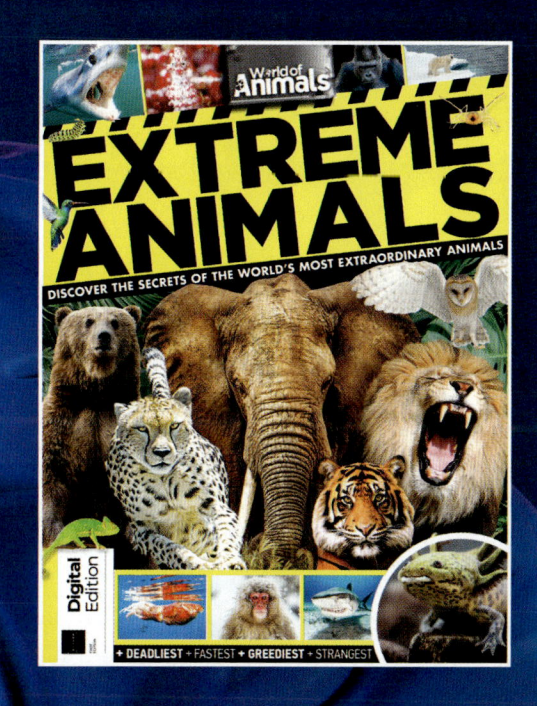

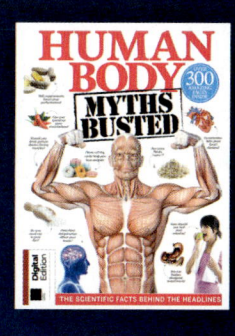

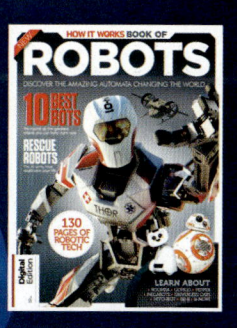

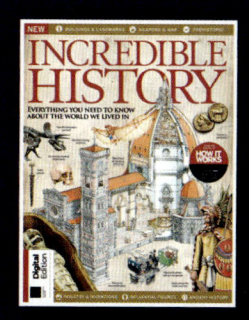

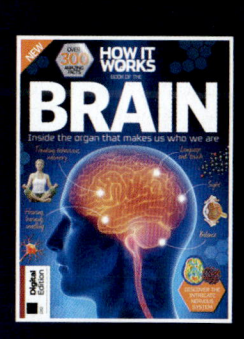

FEED YOUR MIND WITH OUR BOOKAZINES

Explore the secrets of the universe, from the days of the dinosaurs to the miracles of modern science!

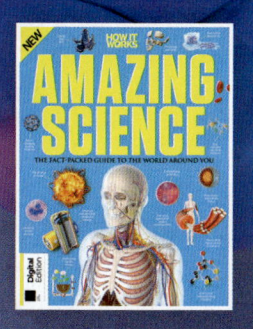

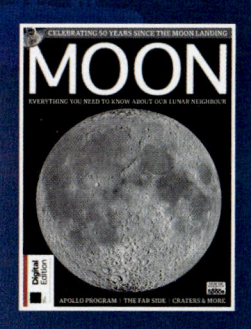

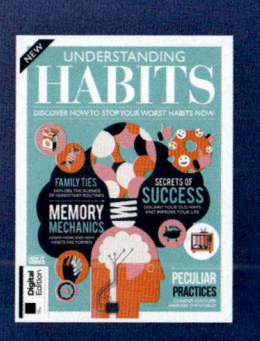

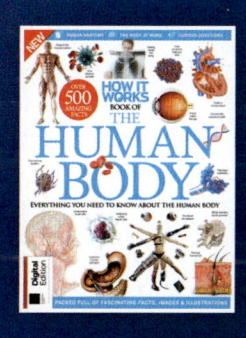

What was Earth like when dinosaurs roamed the planet?

Follow us on Instagram @futurebookazines

www.magazinesdirect.com
Magazines, back issues & bookazines.

FUTURE

SUBSCRIBE & SAVE UP TO 61%

Delivered direct to your door or straight to your device

Choose from over 80 magazines and make great savings off the store price!

Binders, books and back issues also available

Simply visit www.magazinesdirect.com

✓ No hidden costs　🚚 Shipping included in all prices　🌍 We deliver to over 100 countries　🔒 Secure online payment

---

FUTURE

magazinesdirect.com
Official Magazine Subscription Store